PHOTOGRAPHIC REGIONAL
ATLAS OF BONE DISEASE

Second Edition

PHOTOGRAPHIC REGIONAL ATLAS OF BONE DISEASE

A Guide to Pathologic and Normal Variation in the Human Skeleton

By

ROBERT W. MANN, Ph.D.

and

DAVID R. HUNT, Ph.D.

Foreword by

O'Brien C. Smith

Introduction by

Donald J. Ortner

CHARLES C THOMAS • PUBLISHER, LTD.
Springfield • Illinois • U.S.A.

Published and Distributed Throughout the World by

CHARLES C THOMAS • PUBLISHER, LTD.
2600 South First Street
Springfield, Illinois 62704

© 2005 by CHARLES C THOMAS • PUBLISHER, LTD.

ISBN 0-398-07539-5 (hard)
ISBN 0-398-07540-9 (paper)

Library of Congress Catalog Card Number: 2004055376

With THOMAS BOOKS *careful attention is given to all details of manufacturing
and design. It is the Publisher's desire to present books that are satisfactory as to their
physical qualities and artistic possibilities and appropriate for their particular use.*
THOMAS BOOKS *will be true to those laws of quality that assure a good name
and good will.*

Printed in the United States of America
MM-R-3

Library of Congress Cataloging-in-Publication Data

Mann, Robert W., 1949–
    Photographic regional atlas of bone disease : a guide to pathologic and
normal variation in the human skeleton / by Robert W. Mann and David
R. Hunt ; foreword by O'Brian C. Smith ; introduction by Donald J.
Ortner.–2nd ed.
        p. cm.
    Rev. ed. of: Regional atlas of bone disease. c1990.
    Includes bibliographical references and index.
    ISBN 0-398-07539-5 – ISBN 0-398-07540-9 (pbk.)
        1. Bones–Diseases–Atlases. 2. Paleopathology–Atlases. I. Mann, Robert
W., 1949–   . Regional atlas of bone disease. II. Title.

RC930.4.M35 2004
616.7'12–dc22

                                                                        2004055376

FOREWORD

Art is I, Science is We.
—Claude Bernard

Enthusiasm. The most motivating force in a student is enthusiasm. Many bring it with them, already on fire for their particular area of interest. Most though are infected with it by their instructors and fellow students as a sense of discovery, for advancement and competency develops. Enthusiasm dwarfs things petty to science; egos, attitudes, personal agendas, and the like. It creates an aura of academic purity, an environment without fear where "we" is paramount, and "I" becomes a measure of capacity, not importance. It is a time where we can be smart together and we can be dumb together without pride or fear.

Cultivating enthusiasm is one of the hardest tasks for an educator, especially in students just entering an area of study. Many disciplines have their own language, because it requires precisely defined concepts to advance the field. The introductory student needs to acquire some of this to be facile in developing his knowledge and thinking, but too much can intimidate and dampen enthusiasm. The educator, well versed in terminology, needs to introduce his topic in the language of the layman in order to assure communication. This latter is not an easy task because precision of concept suffers.

It is a bold step then for any introductory text to be written especially for the entering student. Colleagues who have already achieved their knowledge-base can always be critical of the authors license and charge oversimplification; and in part will always be right. My reply is that I've rarely found reference books to have a well-thumbed appearance. If I have to choose between precision and enthusiasm for the new student, it will always be enthusiasm! If the fire gets stoked, the opportunity for full potential is achieved.

Let the above be enough to explain this book to my colleagues. I hope too, that they will learn some things from the authors, because I did. For you, the most important reader, the newest generation, I wel-

come you as colleagues and invite you to these pages. Read! Enjoy!
Discover! Think!

O'BRIAN C. SMITH, M.D.
Professor of Pathology;
University of Tennessee School of
Medicine, Knoxville
Chief Medical Examiner;
State of Tennessee

INTRODUCTION

Careful description and classification are basic methodological tools in all categories of science. This is particularly the case in biomedical research where substantial resources are invested in a continuous process of refining diagnostic criteria (classification) for various diseases. The field of paleopathology has been slow in confronting some long-standing problems in description and classification and this has limited its development. The result is a substantial proportion of the existing literature that is of minimal value in clarifying many of the broader questions that must be addressed if paleopathology is to reach its full potential as a legitimate biomedical discipline.

For example, it would be very helpful to have a database that provides information on the antiquity, geographical distribution and evolutionary trends of disease. We also need data that will help to clarify the evolution of complex relationships that exist between the many factors that affect the human response to disease, including: (1) the pathogenic agent, (2) environmental factors (e.g., air pollution) that affect health, (3) nutrition and (4) the immune response of a patient to disease. However, without a clearly defined and generally accepted descriptive terminology and classificatory system it is difficult, if not impossible, to compare the research of one scientist with that of another in building a relevant base of data.

An important contribution to the study of skeletal paleopathology would be the development of a rigorous method to describe the abnormal conditions encountered in archeological human skeletons. It is both surprising and frustrating that after 150 years of research in paleopathology there is still much to do in creating a careful and comprehensive descriptive terminology as well as a general classification of the abnormal conditions encountered in skeletal specimens. Much of how we describe pathological conditions in archeological skeletons is derivative of medical and particularly orthopedic nomenclature and classificatory systems. These systems continue to develop and staying conversant with current usage is a challenging exercise. The major problem, however, is not one of semantics. Rather it is that many of the lesions and their distribution patterns in archeological dry bone specimens bear minimal relationship to descriptive and classificatory features that are central in clinical orthopedic practice. What is crucial for paleopathology is a nomenclature and classificatory system that integrates all of the pathological information that is apparent in skeletal paleopathological specimens. Such a system would necessarily include orthopedic terms and classification where the features were closely related to those used in a clinical setting. There are, however, occasional conditions in paleopathological cases that are not well known in clinical orthopedic practice and a precise classificatory system might demonstrate relationships that previously had not been understood.

In working with both professional colleagues and graduate students I have, for many years, emphasized the importance of first describing carefully what one sees in cases of skeletal paleopathology. Careful description is timeless and, if done well, forever gives future readers of reports the option of reinterpreting your conclusions (i.e., diagnoses). Demographic data, including age and sex, are important factors in interpreting descriptive information. However, the most important element in paleopathological research is the basic description of abnormal bone including the type and distribution pattern within the skeleton. There are four basic abnormalities of bone: (1) abnormal size, (2) abnormal shape, (3) abnormal bone formation and, (4) abnormal bone destruction. There are additional features associated with these general abnormalities that provide helpful supplemental information. For example, is the abnormal bone formation poorly organized (this typically means rapid growth) or well organized (usually slow growth)? Do destructive lesions have well-defined margins with evidence of well-organized bony repair (circumscribed and generally less aggressive) or poorly-defined margins (permeative and generally more aggressive)? These and other features are all critical elements in any interpretation of a paleopathological case of skeletal disease.

The location of lesions within the skeleton provides an important link with clinical experience but one needs to be cautious in making such associations. In dry-bone paleopathological cases one often sees lesions that would not be apparent in clinical radiographs and are thus not well documented in the medical descriptive and classificatory systems. Indeed the added information on skeletal lesions is one of the significant potential contributions that careful study of paleopathological cases can make to a more complete understanding of the skeletal manifestations in orthopedic pathology. A pathology based on dry-bone conditions also means that some distribution patterns of abnormal tissue within a pathological skeleton will vary from patterns established on the basis of radiology in living patients.

Careful description is not easy and I do not wish to underestimate the difficulty of the process. Nevertheless, most people can, with discipline, learn to recognize the essential features of bone reaction to disease. The first step is, of course, a thorough knowledge of normal gross anatomy of bone at all ages from fetal through old age. Archeological skeletal samples are a wonderful source of anatomical knowledge since the entire age spectrum is usually represented. Classification or diagnosis is a much more complicated matter and for many cases encountered by the researcher investigating paleopathology, years of experience and a comprehensive knowledge of orthopedic pathology may be necessary and, even so, may not be possible.

For those conducting research on skeletal paleopathology great attention needs to be paid to using a well defined and widely recognized terminology in describing pathological skeletal lesions. Excellent reference materials in radiology exist such as Resnick's five-volume work (2002). The second edition of my book on human skeletal paleopathology (Ortner 2003) may also be helpful in highlighting both the terminology and the diagnostic options for some cases of skeletal disease.

I am optimistic that further advances will be made in developing descriptive and classificatory methodology for paleopathology in the near future. In the meantime it is important to use descriptive terms and modifiers that are part of the general lexicon we all share. Bone addition, bone destruction, porous bone, and destructive lesions are examples of terms that are descriptive and have wide recognition in many disciplines and I encourage their use. Jargon, is one of the biggest barriers to effective communication that exists and should be eliminated or, at the very least, kept to a minimum. At some point, however, you will need to acquire a working knowledge of medical terminology if only to understand and interpret the existing literature on paleopathology and communicate with medically trained colleagues.

The second edition of the *Regional Atlas of Bone Disease* is a new attempt to assist the beginning skeletal paleopathologist to recognize some of the more common pathological conditions that may be encountered in dry-bone specimens. The authors have provided new cases to illustrate their points but continue to insist that their endeavor be viewed as an initial step in any classificatory process. This is wise counsel, given the complexity of classification. One of the fundamental problems for any classificatory system is that the bone reaction to disease is limited. In view of this it is not surprising that a given pathological condition (i.e., osseous response) may be the result of any one of several pathological processes.

The reader should also be aware of the strengths and weaknesses of a regional approach to skeletal paleopathology. Archeological skeletal samples often do not have complete skeletons. This is particularly true of older museum collections where only the skull and mandible may have been recovered. However, even where an attempt was made to excavate the entire skeleton the result is usually only partially successful. In this context a regional review of pathological conditions may be the only one possible and is certainly helpful. It is also true that many pathological conditions occur in a single location in the skeleton (solitary or unifocal conditions). A regional focus is generally adequate for such lesions.

However, a regional approach is less helpful in multifocal pathological conditions. In this type of skeletal paleopathology, the distribution pattern of abnormal bone is a critical element in classification and the user of a regional approach will need to reconstruct the overall pattern by carefully reviewing the information for each region of the skeleton. A review of the distribution pattern of abnormal bone is important for classification but also contributes to the general understanding of pathogenesis in orthopedic disorders.

Despite this cautionary note, the beginning skeletal paleopathologist should find the new edition of the Regional Atlas a helpful starting point when he or she encounters a skeletal abnormality in archeological burials. Remember, however, first provide a careful and detailed description of the abnormalities you see including the nature of the abnormalities and their location in the skeleton. An attempt at diagnosis can then be made with the assurance that others will at least have the option of reaching a different diagnostic conclusion on the basis of the descriptive information you have provided should that be appropriate. The authors' counsel to seek advice on diagnosis from specialists in skeletal disease is wise. Keep in mind, however, that very few medical specialists have experience with dry-bone specimens and are often as baffled by a pathological case as is the osteologist. The orthopedist does, however, have the advantage of knowing what most of the diagnostic options are and this is a very useful point of departure.

DONALD J. ORTNER, PH.D.
Department of Anthropology
National Museum of Natural History
Smithsonian Institution

ACKNOWLEDGMENTS

The authors would like to extend a special debt of gratitude to Dr. Donald J. Ortner, Dr. Douglas W. Owsley, Mr. Paul S. Sledzik, and Mr. Sean P. Murphy to this enterprise. Each of these scientists played a significant role not only in the development of this book, but its contents. Dr. O'Brian C. Smith contributed substantially to writing the Foreword and Chapter IV. The authors would also like to express their gratitude to Drs. J. Lawrence Angel, Ethne Barnes, William M. Bass, Hugh E. Berryman, Bruce Bradtmiller, Mr. Henry W. Case, Drs. Leslie E. Eisenberg, Eugene Giles, Thomas D. Holland, Lee Jantz, Richard L. Jantz, Marc A. Kelley, Linda Klepinger, Ms. Marilyn R. London, Keith A. Manchester, Marc S. Micozzi, Bruce Ragsdale, Charlotte A. Roberts, T. Dale Stewart, Douglas H. Ubelaker, and P. Willey. It was through their friendship, teaching, and professional guidance that this book came to fruition.

All illustrations were drawn by Robert W. Mann, except Figure 124 (Elizabeth C. Lockett) and Figure 93b (Neil Fallon). Drawings and most of the photographs were based on anatomical specimens at the Smithsonian Institution. Unless stated otherwise, all photographs were taken by the authors. Chip Clark of the National Museum of Natural History, Smithsonian Institution, graciously allowed us to reproduce several of his photographs. The opinions expressed in the Photographic Regional Atlas are the sole responsibility of the authors. Last but certainly not least we thank our wives, Vara Mann and Kim Wells, and our parents Adele D. Mann and Arlys Roehm Hunt and James C. Hunt for their love and support.

CONTENTS

PHOTOGRAPHIC REGIONAL
ATLAS OF BONE DISEASE

Chapter I

USING THE PHOTOGRAPHIC REGIONAL ATLAS

The information contained in the Photographic Regional Atlas–herein referred to as the Regional Atlas–is based on paleopathological examination of more than 8,000 complete or nearly complete skeletons from archaeological sites for forensic investigations throughout the world. The majority of their studies are from prehistoric collections from North America, particularly from the Great Plains, Pacific Coastal regions and the NorthEastern United States; historic cemeteries and burials from Louisiana, Maryland, Nevada, Virginia, Washington, D.C. (including War of 1812, Civil War soldiers and iron coffin burials; as well as approximately two hundred forensic cases including Americans missing in action (MIA). Skeletal collections from Africa, Australia, Central Asia, and South America have also been investigated by the authors. Contemporary skeletal samples have been studied by the authors from the Hamann-Todd (Cleveland, Ohio) and Robert J. Terry Anatomical and George S. Huntington Anatomical (Smithsonian Institution) collections.

The Regional Atlas approaches the recognition of disease according to the bone affected. The format of this handbook begins with a description of how to use the Photographic Regional Atlas (Chapter I), followed by a brief history of paleopathology (Chapter II). Chapter III gives step-by-step instructions on how the authors conduct a paleopathological analysis. Chapter IV briefly covers the mechanics of bone remodeling. The bulk of the Regional Atlas is Chapter V. This chapter deals with specific diseases affecting each bone in the body, beginning with the skull and progressing down the skeleton. Accompanying some lesion descriptions is a statement of the relative frequency (e.g., uncommon to rare finding) or percentage that one might expect to find in most archaeological skeletal samples, in most cases for Native American groups since the majority of the author's studies encompassed these populations.

3

References cited within a sentence indicate that the information was derived from these sources. References at the end of a paragraph (following the period) were included as additional sources for the reader to access for further information. Many of these references are the seminal reports of research for these pathological conditions or anomalies or extensively cover the condition. It is not necessary to reiterate the information published and available in these volumes.

The reader will find that many of the references used in this book were culled from the medical, clinical, and radiological literature rather than the anthropological literature. The reason is multiple. First, clinical studies and case reports provide information based on findings of known age, race, and sex individuals in living groups. Anthropological studies, in comparison, tend to focus on samples of unknown age, race and sex individuals in recent or ancient groups. Second, most diseases can be specifically identified in modern clinical studies, but not necessarily in ancient remains.

Chapter VI provides information on fungal infections. The treponematoses (i.e., syphilis and allied conditions) are summarized in Chapter VII. Chapter VIII briefly discusses tumors, perhaps the most difficult skeletal to be diagnosed. Chapter IX discusses perimortem and postmortem fractures. Chapters VI through IX are designed to only briefly present the effects of these pathological skeletal conditions on the human skeleton. The references cited in those chapters much more extensively cover these particular diseases and should be referred to by the reader for more in-depth research.

As an anatomical overview, dorsal and ventral views of the human skeleton are provided in Chapter X and the major muscle attachments which attach and would be most influential to skeletal morphology are illustrated in Chapter XI.

This book was rewritten after being "field tested" by physicians, students, and paleopathologists for more than a decade. Updated references and findings in the field of archaeology, paleopathology, and medicine have been incorporated into the Photographic Regional Atlas. Most importantly and, as many readers have suggested, the book is spiral bound for easy handling, and most of the drawings have been replaced with photographs to give the reader more detail and a better understanding of exactly what is being presented. This book also reflects the authors' own experiences examining more than 8,000 human skeletons from around the world since the Regional Atlas was first published in 1990. Diseases, lesions, and skeletal anomalies too rare to be expected to be encountered in a routine skeletal analysis have been removed and replaced with those that might be expected to be encountered in most skeletal collections around the world.

It should be remembered that no text could fully or even adequately cover every disease, anomaly, or normal anatomical variant present in the human skeleton; the present text is no exception. While some topics in the Photographic Regional Atlas are discussed in great detail, others are conspicuously brief owing to their extreme difficulty in differential diagnosis or rarity on most skeletal collections (e.g., tumors). One goal of the Regional Atlas was to include the findings and hypotheses of contemporary clinical practitioners (e.g., paleopathologists, radiologists, etc.) to supply the reader with a number of interpretations from which to choose. Such an approach also serves to educate the reader as to the complexity and controversy surrounding the identification, classification, and etiology of many bone diseases.

It is hoped that the experiences of the authors will make it possible for anyone with a sound knowledge of human osteology and skeletal morphology to conduct a basic **descriptive** paleopathological analysis of one or many skeletons. It should be noted, however, that the field of paleopathology is filled with ambiguities and subtleties. Committing this atlas to memory doesn't make one a paleopathologist; only knowledge, training, and above all, experience will qualify you for such a title. The Photographic Regional Atlas will, however, enable you to conduct your own analysis and, in questionable cases, alert you to seek the advice of an experienced paleopathologist, radiologist, or orthopaedist. The importance of a thorough descriptive analysis, however, cannot be overemphasized.

To use the Regional Atlas, first become familiar with what and where lesions, conditions, and anomalies might be expected in the skeleton, locate and identify them in the text, and then refer to the excellent paleopathology, developmental and clinical texts by Allison and Gerszten, 1982; Barnes (1994), Beighton (1978), Brothwell and Sandison (1967), Cockburn and Cockburn (1980), Dieppe et al. (1986), Greenfield (1975), Hauser and DeStefano (1989), Jarcho (1966), Manchester (1983), McCarty (1989), Morse (1969), Moskowitz et al. (1984), Ortner (2003), Ortner and Aufderheide (1991), Ortner and Putschar (1985), Resnick (2002), Resnick and Niwayama (1988), Robbins 1968, Rogers and Waldron 1995, Steinbock (1976), Thijn and Steensma (1990), Tyson and Dyer (1980), Webb (1995), Wells (1964) Zimmerman and Kelley (1982), or other references in the text, especially the *Journal of Bone and Joint Surgery* that deals primarily with the skeleton. Refer also to paleopathology bibiliographies compiled by Crain (1971) and by Elerich and Tyson (1997). While some of these texts may appear to be outdated, they continue to serve the scientific and medical community as some of the most relevant and useable texts in circulation to date. It is hoped that the Photographic Regional

Atlas will serve as a valuable companion to the existing paleopathology literature.

The illustrations in this atlas are predominately specimens from either the National Museum of Natural History (Smithsonian Institution) or the National Museum of Health and Medicine (Armed Forces Institute of Pathology), Washington, D.C. Catalog numbers of the particular specimens will be included where appropriate. And the few other specimens not from these particular institutions will be indicated as to their origin. To avoid continuous lengthy location and collections names, the abbreviations below will be used for identification:

AFIP	National Museum of Health and Medicine, Armed Forces Institute of Pathology, Washington D.C.
NMNH	National Museum of Natural History, Smithsonian Insitution, Washington, D.C.
NMNH-H	NMNH–George Huntington Collection
NMNH-T	NMNH–Robert J. Terry Collection

Chapter II

A BRIEF HISTORY OF PALEOPATHOLOGY

Paleopathology is the study of disease in premodern anthropological and paleontological specimens. Research on this subject dates back nearly two hundred years and has evolved from an area devoted to the study of medical curiosities and exploration of the human body to an integrated discipline of medicine. Johann Esper who, in 1774, reported on a pathological femur of a cave bear from France made one of the earliest studies on bone diseases in dry specimens. Since this earliest study, some scholars have drawn arbitrary temporal divisions marking periods where the primary focus of paleopathology has shifted. For example, before 1900 the focus was on traumatic lesions and syphilis and between 1900 and 1930 infectious disease. Today, the focus is comparative and multidisciplinary with ecology as a major component (Ubelaker 1982).

Some of the more notable early researchers in paleopathology include A. Hrdlicka, D. Brothwell, C.J. Hackett, E.A. Hooton, S. Jarcho, J. Jones, V. Moller-Christensen, R.L. Moodie, M. A. Ruffer, G.E. Smith, R. Virchow, F. Wood Jones, and C. Wells to name a few. Each of these researchers made considerable contributions to the early study of bone disease in humans.

Although some early researchers studied large skeletal populations (e.g., Hooton 1930; Smith 1910), the population approach to disease frequencies was used by only a few, Wyman (1868) and Hrdlicka (1914) being notable examples. Because of the seemingly unlimited field of opportunity and discovery in paleopathology, much of the focus of bone disease was on the more exciting and rare diseases and trauma such as tuberculosis, leprosy, syphilis, trephination, rickets, and the like. Skeletons exhibiting syphilitic-like lesions served to fuel many heated debates and spawned a number of hypotheses concerning its diagnosis in dry bone, as well as its time and place of origin.

While the "exciting" diseases were being debated by some of the world's leading scholars, other researchers around the turn of the twentieth century focused on the less exotic skeletal indicators of health such as cribra orbitalia (Welcker 1888) and symmetric osteoporosis (Hrdlicka 1914). All of the early researchers had to rely on gross examination of the bones because radiographs, refinements in biochemistry, cytology (cell technology), and soft tissue pathology lay before them. Macroscopic examination and a descriptive analysis was the primary method of diagnosing bone disease when the "patient" was an archaeological skeleton with no clinical history. However, this did not stop some paleopathologists from offering their opinions and giving the disease a name (differential diagnosis), including ones that were incorrect. Although the cornerstone of contemporary paleopathological analysis continues to be the descriptive analysis, researchers now have at their disposal extremely sophisticated radiologic, immunologic, and microscopic techniques to aid in diagnosing skeletal lesions.

Contemporary paleopathology has carried on many of the traditions established by its forefathers. For example, much debate still continues on whether syphilis-like lesions in archaeological bones represent venereal syphilis or one of the other treponematoses (e.g., yaws, endemic syphilis); the age-old debate of whether the Europeans spread syphilis to the New World also has yet to be resolved (Akrawi 1949; Baker and Armelagos 1988; Bloch 1908; Brothwell 1970; Cockburn 1961; Dutour et al. 1994, Hackett 1967; Holcomb 1930, 1935; Livingstone, 1991, Pusey 1915; Rothchild and Rothchild 1995, Williams 1932) although innovative immunologic and multidisciplinary research may one day clarify this issue.

While the exotic diseases are still of keen interest to paleopathologists, an added emphasis has emerged focusing on the subtle bony changes that reflect physical stresses and activities of everyday life (e.g., mild indicators of osteoarthritis), nutrition, and diet (Angel 1966, 1976, 1981; Angel et al. 1987; Cook 1979; Cook and Buikstra 1979; Iscan and Kennedy 1989; Mann et al. 1987; Owsley et al. 1987; Schoeninger 1979; Sillen 1981).

Paleopathologists, through the cumulative studies of disease in the human skeleton, now have a better understanding of how many diseases developed and spread both geographically and temporally. The "mundane" skeletal indicators of physical stress, nutrition, and diet are now studied with the same enthusiasm as the exotic diseases. Only by including the full range of bone disease and testing hypotheses are we able to make valid biocultural conclusions on the general and specific health status of skeletal populations. As the late Dr. Larry Angel (1981:513) once wrote, "We are still all amateurs in paleopathology looking toward a bright cooperative future." For a more thorough history of paleopathology refer to Aufderheide and Rodriguez-Martin (1998:1–10); Jarcho (1966); Ortner (2003:8–10); and Ubelaker (1982).

Chapter III

PREPARING FOR A BASIC
PALEOPATHOLOGICAL ANALYSIS

I n preparing to examine a skeleton series it is important to first determine the focus of the study. It is simply not possible to gather every
bit of information on every skeleton. While one researcher might
think it imperative to take bone core samples for lead and other trace
mineral analyses, another may restrict his or her research to nonde-
structive techniques. Prior written permission should always be
obtained before performing any destructive or invasive bone studies.

Before beginning the skeletal analysis you might first like to review,
or at least have at hand, some of the more comprehensive texts on
human osteology including Bass (1995), Brothwell (1981), Krogman
and Iscan (1986), Shipman et al. (1985), Steele and Bramblett (1988),
Stewart (1979), Ubelaker (1999) and White and Folkens (1991). As with
any project, planning and organization are the keys to a successful
skeletal study.

Before going further, we would like to clarify the differences
Between "pathology" and lesion or pathological condition. A "pathology" is not synonymous with a lesion but, rather, the *study* of disease;
a lesion is the response to disease or a wound (Thomas 1985).
Although many learned paleopathologists and clinicians use the words
synonymously, the accurate term for a disease state in bone is lesion,
wound, injury, or pathological condition.

The following suggestions are offered as guidelines on how to pre-
pare for a paleopathological analysis. The technique can be applied to
one or one thousand skeletons and has proven to be comprehensive
enough to satisfy the needs of most researchers. Much of this technique was developed by Drs. Douglas Owsley and Bruce Bradtmiller
and has since been successfully implemented on more than 8,000
skeletons.

First, lay out each skeleton on a separate tray in the correct anatomical position. For example, place the ribs in proper order according to

number and side (e.g., left 1st, 2nd, 3rd, etc.) (see Mann 1993). The vertebrae should also be laid out beginning with the first cervical and ending with the fifth lumbar. The bones of the hands and feet should be separated by side. This method allows comparison of bones from different trays (or archaeological features) for evaluation and matching without commingling the bones.

When working with commingled skeletons (i.e., two or more skeletons mixed together) it is best to take standard 3 x 5 index cards, cut them in half, punch a hole in one corner, affix a rubber band, and attach a card to each bone with a burial or feature number. This method allows bones to be pulled from various trays (or archaeological features) and compared for matches without commingling the remains. Each element can later be put back on its original tray if no matches are made. Tagging the bones also allows them to be put aside for photography and radiography. After fragmentary skeletal elements are reassociated, reconstruction of the elements with glue and/or masking tape can be performed and then inventory and measurements, and a biological profile including the individual's age at death, ancestry (race), sex, and stature can be compiled.

Regarding the topic of reconstructing bones, brief deviation concerning the use of particular adhesives on skeletal remains is appropriate. The choice of adhesive will depend on a number of factors including; whether the skeleton will be subject to additional forensic investigations where the introduction of adhesive chemicals may be detrimental to other tests, such as DNA, trace element and isotopic analysis. In archeological contexts, questions of whether the skeletons will be reburied, how long they will be available for study, will the bone be dated by C-14, and is the reconstruction planned for long-term curation (Greta Hansen, pers. comm. 1989). In the latter questions, the stability of the adhesive is of interest. Although Duco™ cement has been widely used for reconstructing bones, in time it becomes dry, yellow, brittle, and separates at the glued "joints." The inherent instability of Duco cement renders it unsuitable for long-term use. If the reconstruction is meant to be permanent, a better choice is Acryloid B-72 or B-48N (available from Rohm and Haas) or solvented polyvinyl acetate (PVA) (available from Union Carbide Corporation). For more information on reconstructing, preserving, and consolidating bone, consult an objects conservator or refer to Johansson (1987) or Sease (1987).

The next step in conducting a skeletal examination is to examine each bone for pathological conditions. The most important technique in the analysis is examining each bone closely, using a standard fluorescent lamp. Hold the bone approximately 3–5 inches from the lamp and carefully inspect the surface inch by inch. Do not pick up a bone and quickly examine it for obvious pathological changes since many subtle lesions and, possibly, cut marks from scalping (Neumann 1940; Ortner 2003; Steinbock 1976) or defleshing (Ubelaker 1999) may be overlooked. It's a good idea to examine each bone by holding it in different positions to change the angle of the light as it strikes the bone. This, too, may reveal a small lesion or trauma that was formerly hidden by shadows. It should be stressed that the use of a hand magnifying lens or an illuminating magnifying lamp are necessary to observe subtle features on or in the lesion or defect. If you find yourself staring at a questionable lesion, put the bone aside and return to it later. You might use this time to reference other paleopathology texts to help clarify your questions. Being consistent with your scoring criteria (e.g., mild) for judgment is critical.

If, after examining ten or fifteen skeletons, you find that you have been scoring a lesion as moderate in severity and feel that it should only be scored as slight, then you should modify the criteria for the trait or lesion in question before going further. Obviously, it's easier to go back and change data sheets for ten or twenty skeletons than to realize this after you've analyzed one hundred or more. At such a late point in your examination, you might find that you've run out of time or are too frustrated to reexamine the skeletons. Be observant early in your analysis and note those conditions that are ambiguous in either severity or presence/absence. After the skeletons have been examined for pathological conditions, bones needing to be X-rayed and/or photographed should be clearly labeled and set aside.

Note at the top of each inventory sheet and notes (separate pages for each individual) the burial number, age, sex, and any unusual features present in the bone, or associated with the particular burial (e.g., "projectile point in 3rd lumbar. Photo/X-ray"). This information will later save you time. Also ensure that you put the examiner's name and date of the research on each page for future researchers to reference.

Finally, if you have access to a laptop computer you can save many hours of deciphering, reexamining and rewriting laboratory or field notes for each burial. This technique is especially important in conducting a thorough **descriptive** paleopathology analysis. If you use the computer as a source of recording your methods, findings, and thoughts, a large part of the write-up will already be done; this "rough draft" can then be edited and will produce a very detailed report. (You'd be surprised how much information is forgotten after the analysis is completed.) The following is an example of how a skeleton might be described using a laptop computer:

Skeleton 1. Present is the poorly preserved skeleton of a young adult male, perhaps 20–25 years of age. Present are the skull, mandible (missing right ascending ramus), both femora, tibiae, left innominate, right foot, and left clavicle.

Age is based on examination of the pubic symphysis (billowy), auricular surface of the ilium (no microporosity), dental development and eruption, cranial and maxillary suture closure (the incisive shows no obliteration), and medial epiphysis of the clavicle (early stage of formation).

Sex is male based on the morphology of the innominate (very narrow sciatic notch and triangular pubic bones), robust skull (well-developed nuchal crest and large mastoids), and femoral head diameter measuring 54 mm.

Pathological conditions. There is an ovoid, healed depressed fracture in the frontal bone immediately above the right orbit. Fracture lines extend from the center of the injury and radiate into the superior orbital plate. There is no evidence of healing, which suggests that this injury may have occurred at or near the time of death (antemortem). The inner vault of the skull was not affected, as there are no fractures or depressed bone visible. The shape of the injury suggests that an oblong object struck this individual above the right eye. Examination of the remainder of the skeleton reveals no further evidence of trauma (it should be noted, however, that many bones are missing).

The distal right femur exhibits a small area of healed periostosis along the lateral surface. There is slight porosity (osteoarthritis, or OA) in both temporal fossae (temporomandibular joints) and the right mandibular condyle is slightly flattened. The teeth show moderate wear, two periapical abscesses of the right mandibular first and second molars, and slight linear enamel hypoplasia.

Comments. The distal left radius exhibits a small area of green copper salts staining (about the size of a quarter) on its posterior surface (need to photo). The location of the stain suggests that a bracelet or other metallic artifact was in contact with the left wrist at the time of burial. The field notes do not, however, report any artifacts recovered in association with the burial.

Chapter IV

FUNDAMENTALS OF BONE
FORMATION AND REMODELING

The purpose of this chapter is to give a brief account of normal and pathological features of remodeled bone in the human skeleton. With a core knowledge of normal bone growth and remodeling, abnormal responses of bone can be appreciated by identifying any reactive changes in gross appearance. The full developmental sequence of the skeleton is left to the domain of the embryologist and pediatrician. Here basic facts are presented, sufficient to impart an understanding of bone formation and remodeling. The reader can also refer to an excellent overview on bone biology, including cell proliferation, types of bone cells and chemistry, growth and development and remoldeling by Ortner and Turner-Walker (2003). See also bone biology and articular joint physiology overviews in Shipman et al. (1985:18–77) and Steinbock (1976:3–16).

THE NECESSITY OF KNOWING

Recognizing a bone as abnormal raises questions about the cause of event (etiology) resulting in remodeling. These findings may be of forensic significance for the individual or reflect the culture of the population. Certain fracture patterns (parry fractures) may tell us of the warlike nature of the people or indicate the introduction of the horse (femoral/pelvic fractures). The ability to recognize infection (e.g., tuberculosis) helps us to understand the epidemiology of certain diseases and reflects hygiene and social activities or public health practices. Other changes may be characteristic of nutritional deficits (e.g., rickets (Mankin 1974a&b, 1990), genetic predisposition to deformity (congenital) or tumorous lesions. But before any ability in recognizing disease can be acquired, a confident practical knowledge of what is normal is a must.

Functionally the skeleton supports the body, protects internal organs, and serves as attachment for the muscles and soft tissues. Bone is a living tissue consisting of 92 percent mineral or solids and 8 percent water. The solid matter is chiefly collagen matrix hardened by impregnation with calcium salts (Thomas 1985). Bones develop either from small cartilage models (anlage) in the eight-week-old embryo or from condensed embryonic tissue (mesenchyme) that forms a dense membrane (Arey 1966). Facial bones and portions of calvaria, mandibles, and clavicles derive from the latter and are called "membranous," all other bones form in areas occupied by cartilage, which they gradually replace and are therefore "cartilaginous."

Bone forms along the paths of invading blood vessels. Chondrocytes enlarge (hypertrophy) and proliferate (hyperplasia) about the blood vessels, becoming osteocytes as the cartilaginous matrix becomes mineralized. Enchondral ossification is a well-ordered sequential process of converting the cartilaginous model into bone. It is present under the perichondrium, a blood vessel rich layer of cells outside the model, and in active centers that develop within the bone. The fewer bones of membranous origin form along the many blood vessels and develop within the membrane. Both of the processes are referred to as *modeling* and any subsequent changes requiring resorption of preexistent bone followed by deposition of new bone is remodeling. Thus modeling is an early process, while *remodeling* occurs during normal growth and continues until death.

All bone is remodeled along a blood vessel advancing through cartilage or bone. Just ahead of this blood vessel is a cutting core of osteoclastic activity dissolving the bone. New bone is formed by the osteoblasts trailing beside the vessel. This is often referred to as a bone-remodeling unit (Rockwood 1989).

Osteoblasts form the bone matrix osteocytes, serve to maintain it, and osteoclasts remove it. The ratio of osteoblasts to osteoclasts determines whether bone is deposited or resorbed. Since a single osteoclast destroys in 36 hours the same amount of bone that ten osteoblasts produce in ten days, it is obvious that if microscopic examination reveals an abundance of osteoclasts, bone resorption must be occurring (Snapper 1957).

Seven basic categories of disease may affect any sort of tissue or organ and are best recalled by use of the mnemonics KITTENS:

> K = Congenital/Genetic
> I = Inflammatory/Infections/Idiopathic/Iatrogenic
> T = Traumatic
> T = Toxin

E = Endocrine/Metabolic
N = Neoplastic/Neuro-mechanical
S = Systemic

Congenital may refer to structural anomalies at birth, or genetic defects appearing later such as osteogenesis imperfecta. Inflammation of any sort, especially chronic infection, frequently causes remodeling. These changes may also be due to unknown causes or a poorly understood disease (idiopathic osteoarthritis) or caused by medical intervention (iatrogenic). Trauma is usually followed by the osteoblastic reparative process after initial resorption at the site of hemorraghe. Toxins may interfere with bone growth (lead poisoning) or become incorporated in the bone (e.g., lead, arsenic, tetracycline). Endocrine or metabolic effects vary markedly, from excessive growth of immature bones (gigantism) or the mature skeleton (acromegaly). Metabolic effects are most pronounced in nutritional deficiency states—vitamins (rickets, scurvy) and calcium deficiency (osteoporosis). Neoplasia may present as areas of resorption, with or without reactive deposition to produce a sclerotic margin. Neuro-mechanical defects such as tabes doralis from tertiary syphilis or the loss of sensation from diabetes mellitus destroy the control of smooth joint action producing traumatized points from the stressful gait. Systemic diseases like rheumatoid arthritis and lupus produce inflamed joints with recognizable patterns of destruction (see Lawrence 1970a; Mankin 1974a&b, 1990; Rogers and Waldron 1995:55–63 for a further discussion).

Osteoarthritis, however, is an example of a widespread inflammatory disease that is poorly understood and can be thought of as systemic (degeneration accompanying increased age) or traumatic in origin (Mankin 1968, 1982). The important point is that any classification system is not rigid but serves as a basis for organizing thought. Another factor to keep in mind as one holds the dry bone in hand is that extraordinary changes may not be due to a primary disease of bone, but the bone has been affected secondarily in the natural course of a disease, as in general inflammatory responses, necrotic results of effected blood flow as in sickle cell and hemophilia. Refer to Raisz (1999) for discussion on the metabolic changes in bone remodeling and the pathological changes which produce osteoporosis and bone loss.

Description of the lesion is important in developing a list of differential diagnosis. Normal bone has a continual balance of resorption and formation and is subject to gradual remodeling of trabeculae to meet changes in stress loading. When stimulated, this balance is upset, but bone can react in only three ways (resorb, deposit or both).

Recognizing the *predominant* process is the key to description. Once the pattern has been described by location, gross, and radiographic appearance, the number of disease possibilities can be narrowed.

For example, a single resorptive (lytic) lesion of a vertebral body could represent infection (e.g., tuberculosis), a primary cancer (plasmacytoma), or a benign lesion (Schmorl's node). If the lytic margin is surrounded by a reactive growth of bone the process is probably a chronic infection (e.g., tuberculosis), however, there is also osteoblastic response around metastises from prostate and breast cancer. Aggressive lysis with no reaction favors malignant disease (such as those of renal, lung and thyroid metastises), while a smooth regular margin is probably both benign and quiescent (Coley 1960). Resorption of bone can also be caused by contact pressure erosion from tumors or along the path of dilated blood vessels. The latter occurs when an alternate pathway must be used after normal circulation is obstructed by congenital or acquired disease.

Deposition of bone is most frequently due to inflammation of the periosteum. This is most readily seen in the exuberant callus after a fracture, but it is also seen in more subtle injuries where muscular injuries next to the bone lift the periosteum away from the cortical bone and blood collects underneath the tissue. The blood clot is organized into scar tissue and may be converted into bone. Chronic infections (osteomyelitis) of bone rarely heal spontaneously, producing years of inflammation resulting in very large deposits of bone, often with a central sinus (cloaca), which continually drained pus. Tumors, especially primary bone cancers, produce wildly aberrant patterns of new bone admixed with areas of aggressive lysis. Hematologic disorders (thalassemia, sickle cell) are associated with exuberent bone proliferation causing radical enlargement of the diploe in response to the formation of additional marrow space in compensation to the anemia. The trabeculae become vertically oriented and produce a "hair on end" appearance in radiographs.

The above examples merely touch on the range and diversity of resorption, deposition, or both, each requiring interpretation of its pattern, speed and interaction for accurate diagnosis.

OSTEOARTHRITIS

A chapter on bone remodeling is not complete without a discussion of osteoarthritis (OA), also known as degenerative joint disease (DJD). Not only is OA chiefly recognized by remodeling changes in skeletal series, it is the most common manifestation of disease after dental caries. It is present in most persons older than 50 years, and in 90 percent of octogenarians, although pathological and biomechanical studies have shown that age alone is not a sufficient cause of OA, as there are enormous differences between the old and the arthritic joint (Bayliss 1991; Lasater and Groer 2000; Watt and Dieppe 1990). Alexander 1990; Felson et al. 1987; Resnick and Niwayama 1988.

As widespread as it may be, the causes of this disease are diverse and obscure. Osteoarthritis most frequently results in destruction of weight-bearing joints (vertebrae, hips, knees) although any joint may be affected. Two clinical patterns emerge, primary and secondary OA. Primary OA is idiopathic, of no known cause, but may be one of the many changes of aging. The spine is most often affected but the patient largely remains unaware of the process. Secondary OA is related to "wear and tear." Traumatized joints (including repetitive micro trauma), congenital abnormalities, and other etiologies (remember KITTENS), are subject to the inflammation and repair process of any living, vascular tissues. Thus, the remodeling is limited to a specific degenerative process and hence the alternate term "degenerative joint disease."

A theory has been advanced to explain each clinical pattern. The biochemical theory seeks to explain primary OA as an aging phenomenon where the body gradually loses its ability to maintain joint cartilage. Stresses develop, damage the joint and the attendant inflammation in the synovium (synovitis) releases enzymes and inflammatory chemicals that attack the cartilage (Mankin 1967, 1968). Once the cartilage is damaged, the changes common to both clinical patterns follow and are irreversible.

A biomechanical theory emphasizes loading stresses directly injuring the cartilage cells, affecting their ability to maintain the matrix. The cartilage becomes soft, with fissures cracking the surface and flakes breaking off. The inflammation induced by this damage then chemically attacks the remaining cartilage and a cycle of accelerating damage develops. Changes in the underlying bone may accompany the process leading to microcysts beneath the cartilage, which then collapses from loss of support (Mankin 1967, 1974, 1982).

Regardless of the mechanism, the final appearance is common to both. With the loss of chondrocytes and fragmentation of the softened cartilage, the subchondral bone becomes exposed. Exposed bone forms small callus rich blood vessels followed by an extensive remodeling process. Without cartilage or a synovial sack to protect the two bony surfaces from abrading, "bone on bone" wear produces a thick polished (eburnated) bone resembling ivory. Other areas show osteoclastic activity with microcysts and even small fractures. The cartilage spreading out at the joint margins and the synovial lining turn into ossified outgrowths known as spurs (osteophytes), indicative of proliferative osteoarthritis, as opposed to erosive changes (for example, periarticular cysts and surface pitting). Chips of cartilage or broken bone spurs may float freely in the joint, called "joint mice" (see Figure 206).

The remodeled subchondral bone and osteophytes are obvious changes typical of OA. But hidden in the marrow cavity is another change. The trabecular pattern, which identifies the lines of stress, differs from the normal pattern and reflects changes in the stress loading of the bone. There is some debate as to whether this reflects cause or effects of the OA. Whichever it is, radiographs soon reveal its existence.

Radiological and clinical features of OA in the living are: (1) narrowing of the joint space, (2) osteophytes, (3) altered bone contour, (4) subchondral bony sclerosis and cysts, (5) periarticular calcification, and (6) soft tissue swelling (Dieppe et al. 1986). Most of these changes are helpful to the anthropologist. Questions anthropologists raise are primarily concerning the predictive value of the presence of OA in skeletal remains. Specifically, does OA reflect changes due to continuous stress or trauma to a joint induced by characteristic activities ("overuse" hypothesis; Alexander 1990)? What are normal ranges and the limits of the normal "wear and tear" for old age? Can differences be found between the dominant and nondominant hands? These questions are particularly pertinent as physical anthropologists examine remains, devoid of soft tissue and lacking medical histories, trying to interpret what activities resulted in joint lesions as proposed by Kennedy (1998). As the reader will see, no statement will be made that these questions can be adequately answered. We can only supply the reader with findings of several clinical studies focusing on the etiology and frequency of OA of contemporary groups.

A number of groups have reported comparing the incidence of OA in various occupations. For example, increased frequencies of OA were found in bus drivers and cotton pickers (Lawrence 1961, 1969), foundry workers (Mintz and Fraga 1973), persons (porters) carrying heavy loads on their heads (Jager et al. 1997), print setters (from flicking letters across the tray with their thumbs, Dieppe et al. 1986), and coal miners (Kellgren and Lawrence 1958). However, pneumatic hammer operators (Burke and Fear 1977), long distance runners (Puranen et al. 1975), soccer players (Adams 1979; Klunder and Hansen 1980), and parachutists (Murray-Leslie et al. 1977) that should expect to experience increased frequencies of OA, don't. That physical activity is clearly one path to recovery and increased health and mobility is evident in that most rheumatologists endorse physical exercise for managing both degenerative and inflammatory joint disease (Burry 1987). Despite the voluminous research and medical attention given it, OA remains a topic of controversy and uncertainty.

A study of 134 individuals aged 53–75 from Northern California, for example, addresses the hypotheses that increased frequencies of OA are associated with the dominant hand (93% were right-handed) or due to "wear and tear" (Lane et al. 1989). The subjects were studied using radiographs, rheumatologic evaluation, and questionnaire. Ninety-five percent of the individuals were classified as having occupations that were nonphysical. The subjects were separated into subgroups based on lifetime heavy hand use. The researchers found ". . . no significant differences or meaningful trends in any subgroup between dominant and nondominant hands" (Lane et al. 1989). However, the authors concluded that since the subjects were in nonphysical occupations they might not have had enough chronic stress or trauma to accelerate joint degeneration in the dominant hand. This study as well as others showed that heavy use remains an open question. A broad range of wear may result in no adverse effects to the joints or the development of OA (Lane et al. 1989; Panush et al. 1986; Wright 1980). Thus, interpretations of skeletal remains deserve no less caution.

This "overuse" hypothesis, widely held by anthropologists and clinicians alike, appears to be rooted in the belief that "too much physical activity" causes osteoarthritis. Following this logic one would conclude that oarsmen, toolmakers, blacksmiths, sharpshooters, and archers would develop osteoarthritis (not just soft tissue inflammation) from overusing their joints in these activities. This is an especially attractive way of thinking when reconstructing past behaviors in ancient groups through skeletal analysis. Interestingly, studies of professional keyboard and string musicians have failed to reveal increased incidences of OA in these activities practiced over long periods of time involving substantial forces on vulnerable joints (Kellgren et al. 1980; van Saase et al. 1989; Radin et al. 1971). The senior author has long believed that the "overuse" hypothesis is too often and too liberally used as an explanation for frequencies/patterns of osteoarthritis and activities leading to its development in ancient groups. It's likely that OA is a result of genetic and environmental variables rather than activity or "overuse" alone.

Readers wishing more detailed information on bone disease and remodeling should refer to Crelin (1981), Manchester (1983), Ortner and Putschar (1985), Ortner (2003), Steinbock (1976), or one of the many texts on embryology and pathophysiology.

Chapter V

DISEASE OF INDIVIDUAL BONES

SKULL AND MANDIBLE

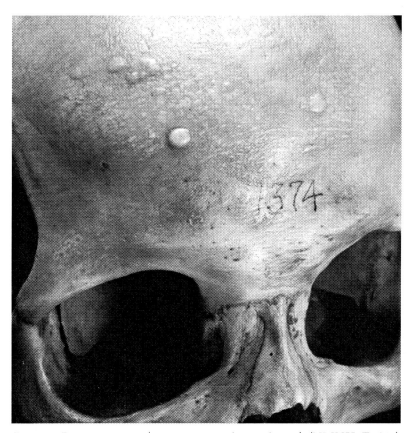

Figure 1. Button osteoma (ivory osteoma; button lesion). (NMNH–T 1374)

S mall to large (ca. 1 cm), polished and roughly circular raised area(s) of dense bone that resembles a small mound or dome–these growths are classified as benign (harmless) tumors but, in fact, are not true tumors, but benign condensations of bone that, according to Eshed et al. (2002), better fits a hamartoma (button lesion) than an osteoma or exostosis suggesting an evolutionary background (**Figure 1**). A common occurrence in most populations (37.6% in modern groups and 41.1% in archaeological specimens; Eshed et al. 2002. Brothwell 1967; Bullough 1965; Bushan et al. 1987; Grainger et al. 2001; Ortner and Putschar 1985; Steinbock 1976.

Small (0.5 mm) to large (2.0 mm) sieve-like holes involving the outer table and diploe accompanied by increased vault thickness–there is considerable debate over the etiology and proper classification of this condition (**Figure 2**). Some populations will show an extremely high frequency of porotic hyperostosis (e.g., coastal Peruvian), while other groups show little or none (e.g., contemporary American whites and blacks). Hrdlicka (1914), in examining 4800 crania, found porotic hyperostosis to be common among the prehistoric coastal peoples of Peru but absent in groups from the mountainous areas.

Regardless of the proposed etiology (e.g., iron deficiency or hereditary anemia (Miles 1975; Salvadei et al. 2001), nutrient losses associated with diarrheal disease (Walker 1986), vitamin or nutritional deficiencies (McKern and Stewart 1957), toxic causes (Hrdlicka 1914), or infection, it is imperative that an accurate descriptive analysis be conducted. As the name implies, porotic hyperostosis should only be applied to those cranial bones that exhibit both porosis (holes) and thickened (increased) bone. Ortner (2003:383–393) identifies a pattern of porosity on the parietals, frontal and especially porosity on the sphenoid, ascending ramus of the mandible and on the superior regions of the scapula as an indicator of scurvy. Porotic hyperostosis should not, however, be confused with hyperplastic conditions (Figure 52) or cancers that do not fit the geographical distribution on the skeleton or appearance of porotic hyperostosis.

Many crania will show tiny pits (porosity) of the parietals (most commonly), occipital, and frontal bone near bregma; however, no thickened bone will be present. Since porotic hyperostosis must by definition include the presence of thickened bone, the present authors have chosen to use the purely descriptive term ectocranial porosis (**Figures 3, 4, 6**) for pitting of the outer vault giving it an "orange-peel" texture, not accompanied by thickened bone. Some researchers (Carlson et al. 1974; Lallo et al. 1977) believe that cribra orbitalia (**Figure 13**) is an early form of porotic hyperostosis. Angel 1964; Carlson et al. 1974; Cybulski 1977; Dallman et al. 1980; El-Najjar and Robertson 1976; El-Najjar et al. 1975; El-Najjar et al. 1976; Hill 1985; Keenleyside 1998; Lallo et al. 1977; Lanzkowsky 1968; Mensforth et al. 1978; Ortner 2003:363–382; Palkovich 1987; Papadopoulos 1977; Ponec and Resnick 1984; Rothschild 2001; Steinbock 1976:213–252; Stuart-Macadam 1985, 1987a, 1987b, 1989, 1992; Walker 1986.

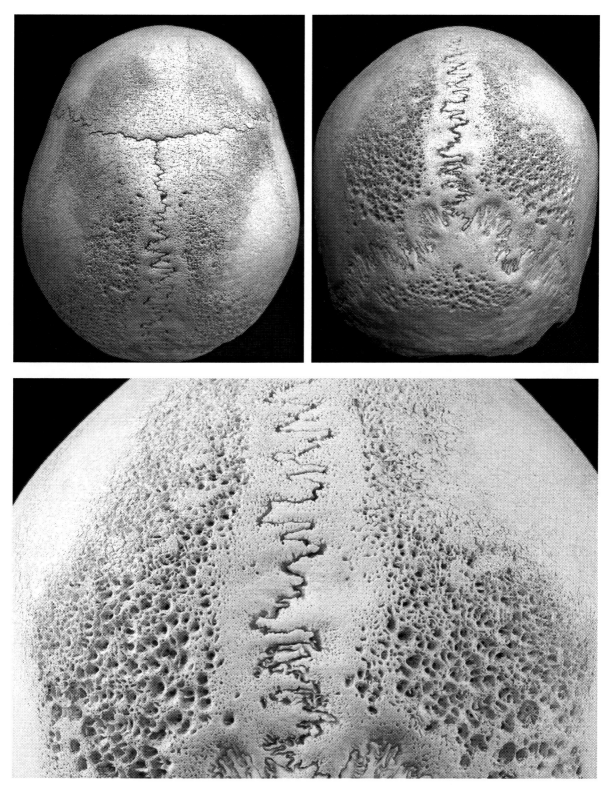

Figure 2a, b & c. Porotic hyperostosis (Angel 1966; El-Najjar et al. 1976; Lallo et al. 1977; Palkovich 1987; Schultz 2001), symmetrical osteoporosis (Zaino, 1967), symmetric osteoporosis (Hrdlicka 1914). (NMNH 264543)

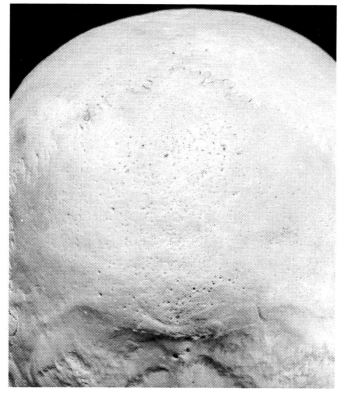

Figure 3. Ectocranial porosis on the occipital. (NMNH)

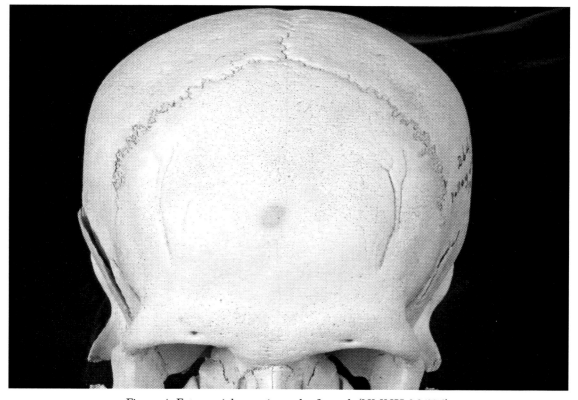

Figure 4. Ectocranial porosis on the frontal. (NMNH 264776)

Tiny pits in the outer surface of the vault that are not accompanied with thickened bone. Although it is difficult to distinguish what the senior author refers to as "ectocranial porosis" from conditions reflecting anemia (porotic hyperostosis/PH), great care must be exercised when identifying outer vault porosity reflecting PH. One way to avoid having to choose an interpretive "diagnosis" is to carefully describe the location, appearance and distribution of pitting in the outer vault. Many "normal" crania exhibit pinpoint porosities in the frontal, parietals and occipital bone. Common occurance, most commonly seen in mid-aged adults.

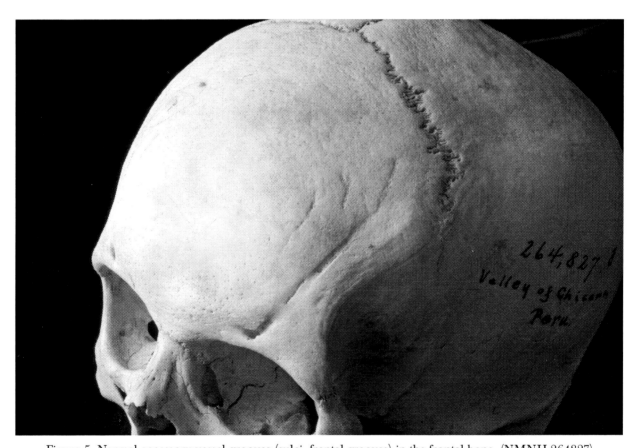

Figure 5. Normal accessory vessel grooves (sulci, frontal grooves) in the frontal bone. (NMNH 264827)

One or more grooves above the orbits for transmission of branches of the supraorbital vessels and nerves (Grant 1948) (**Figure 5**). Sometimes these shallow grooves trail into the supraorbital notch or foramen (see also **Figure 4**). This is considered a normal variant (nonmetric trait). See variants presented in Hauser and DeStefano (1989: 48–50, Plate VII) for different degrees of expression. Rhine (1990) attributes this feature to be more frequently observed in American Blacks.

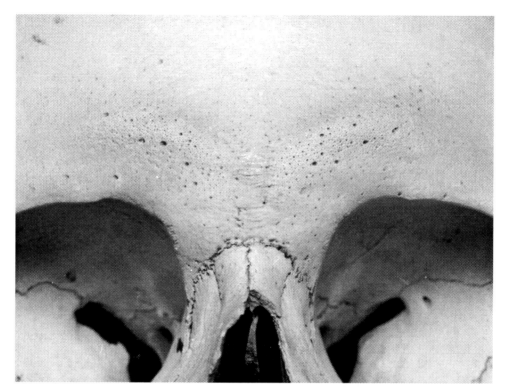

Figure 6. Normal pitting/pinpoint porosity in the frontal bone, often more pronounced in males. (NMNH 264417)

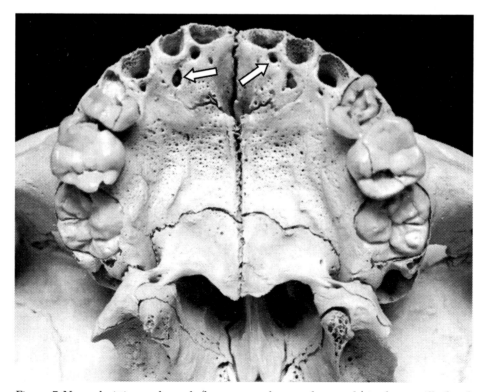

Figure 7. Normal pitting and canals (known as gubernacular canals) in the maxilla/hard palate of a child. (NMNH)

Children's palates are usually more porous and pitted than adults. Children also have multiple canals in the anterior portion of the maxilla due to normal growth and development of the teeth (some of these canals are remnants of dental crypts, while others will guide the permanent teeth through the palate) (**Figure 7**).

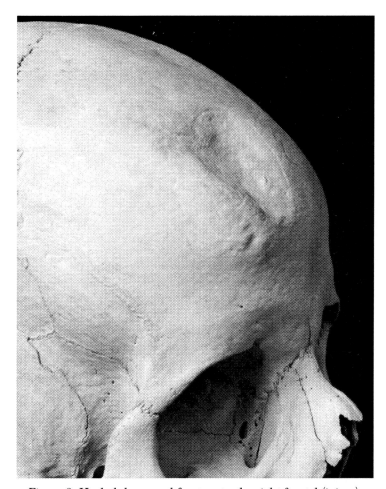

Figure 8. Healed depressed fracture to the right frontal (injury).

Such fractures may be found in any area of the skull but are most frequently seen in the frontal and posterior or lateral portions of the parietals (**Figures 8 & 9**). The typical appearance of a depressed fracture is a concave defect in the outer vault, with or without radiating fractures. The size of the wound is usually about the size of a dime or nickel and circular or ellipsoidal, although any size and shape may be encountered. Often it is difficult to distinguish a depressed wound from a healed lesion originating in the scalp. Depressed cranial fractures, such as in this example, often result in a "pond" fracture because

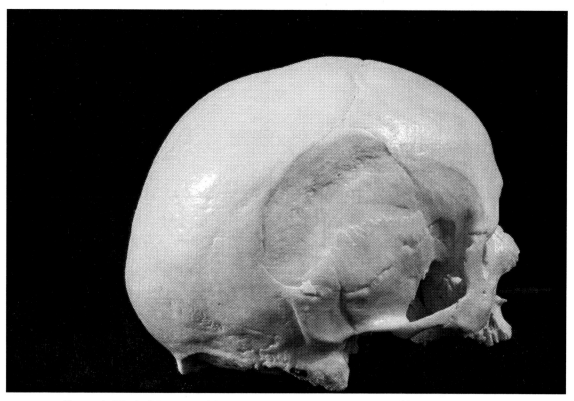

Figure 9. Healed Depressed fracture to the right lateral parietal (NMNH–H 321498)

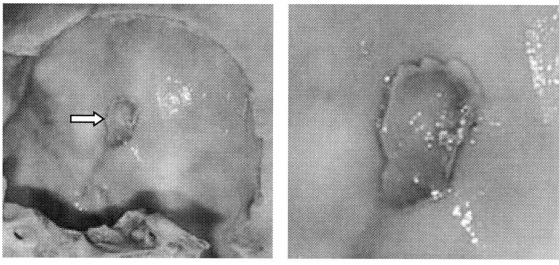

Figure 10a&b. Blunt force injuries to the skull resulting in concentric fractures to the right temporal and parietal (left photo) and an "island" of bone (arrow) displaced but adhering to the endocranium opposite the area of impact in the left parietal.

of its shape. Frequent finding in many groups, especially ancient American Indians (Walker 1989).

Depressed fractures result from a variety of weapons and conditions. Although a large nonhealed (perimortem) fracture of the skull does indicate trauma to the head, it usually will not leave any evidence of the specific type of weapon used or the circumstances of the traumatic event(s). Further, healed depressed fractures may cause the investigator to suspect that an assailant struck the individual, although this may not be an accurate interpretation. For example, the person might have fallen from a rock ledge, struck his or her head, only to die a week later. The fracture pattern could be identical if the person had been struck over the head with a blunt instrument (e.g., war club) resulting in immediate death. Obviously, a projectile point embedded in bone leaves no doubt that some form of conflict occurred. What is important is that bony trauma, whether the result of blunt force or sharp force (bladed or edged implement) must be carefully examined, described, photographed, and measured. The key phrase is "described and documented" with drawings and, preferably, photographs with scale. Many times, speculative interpretation is best not attempted. Be sure to examine the endocranial surface of the skull, as a blow to the outer vault often results in radiating fractures and/or displacement of a portion of bone endocranially. It might also be helpful to hold the skull up against a bright (100 watt or greater) desk light so that the light shines through the bone, illuminating any unusually thin areas, fractures, or displaced bone (**Figure 10a & b**).

Another important aspect of trauma is the pattern of the wound(s) in the population under study. For example, are all of the depressed fractures round? Where are the wounds located? How many sites of trauma show healing suggesting that the victim lived for some time after the event? Again, the overall population picture is important in reconstructing the events surrounding the trauma (Jurmain 2001).

Depressed fracture with no healing (perimortem)–look closely for any indication of healing has occurred, there will be small patches of periosteal new bone growth and possibly porosity in the area of the defect, or along the fracture lines. If death occurred immediately or soon after sustaining trauma (up to about two weeks), the bone may show no new growth. As can be seen in **Figure 9**, any number of objects or implements could have been responsible for the shape of the defect. The shape of the wound would depend, for example, on whether the person was hit with the pointed end of a club (small, circular depressed wound), or the blunt end (larger, oval or oblong depressed wound). Again, the pattern of the wounds in the population under study may help to clarify the type of weapon used. Research by

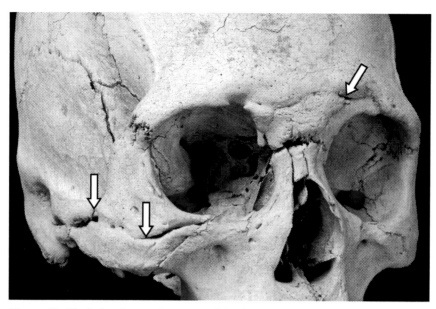

Figure 11. Healed radiating fractures of the frontal, malar and zygomatic arch. Healing is evidenced by the presence of several bridges of bone that span the fracture lines. (NMNH 379259)

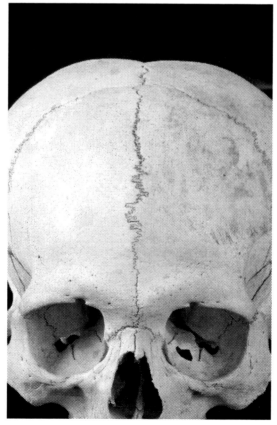

Figure 12. Metopic suture (normal variant). In children before the age of two, the metopic suture normally separates the two frontal bones down the midline. (NMNH 264597)

one of the authors on documented cases of skeletal trauma suggests that the term perimortem, when based solely on the amount of healing visible in bone, must be used with caution, as osseous remodeling may not be grossly discernible until 10 days after sustaining trauma. The rate of healing, however, is heavily dependent on age such that most rapid healing occurs in children and slows with increased age.

One of the authors (RWM) has found it very useful to shade in the shape of a skull wound using a piece of paper and pencil or artist's charcoal. If the depressed wound is in the frontal bone it would be "shaded" using the following method: make a drawing (approximate anatomical size) of the frontal view of the skull, note the approximate position of the wound, and correspond this with the drawing. Hold the drawing against the frontal bone and use the side of the lead to rub across the defect. The result is an unshaded (white) area above the depressed area (the lead won't come in contact with the concave areas). This method renders a silhouette of the size and, more importantly, the shape of the wound. You then can compare the depressed wounds in all of the skulls to see if a size and shape pattern is present which may reflect the weapon(s) used.

This suture (**Figure 12**) usually begins to obliterate at the end of the first year after birth with fusion being complete not later than the fourth to sixth year (Limson 1924). In some individuals (1% to 12% of adults. Krogman and Iscan 1986), the two frontals fail to unite resulting in "metopism" or a persistent metopic suture (extending from nasion to bregma). Remnants of the metopic suture in the form of transverse irregular fissures above nasion are common findings while metopism is uncommon to common. Barnes 1994:148–152; Chopra 1957; Harbert and Desai 1985; Hauser and DeStefano 1989:41–44; Hess 1945; Latham and Burston 1966; Manzanares et al. 1988; Schultz 1918; Torgersen 1950, 1951.

This condition is usually bilateral and appears as small to large holes in the upper surfaces of the orbits (**Figure 13**). In children, the bone may actually be thickened and spongy-like while in adults only remnants of the holes (frequently only pits) remain. The frequency of this condition varies greatly by population and depends on a number of factors, many of which are under debate (e.g., iron-deficiency anemia, perhaps related to malnutrition, scurvy, chronic gastrointestinal bleeding, ancylostomiasis, and epidemic disease. Hirata 1988). Cribra orbitalia may or may not be accompanied by pitting and/or thickening of the outer table of the skull. Stuart-Macadam (1989) suggested "The similarity between porotic hyperostosis of orbit and vault with respect to macroscopic, microscopic, radiographic, and demographic features supports the idea of their relationship." Carlson et al. 1974; Cybulski

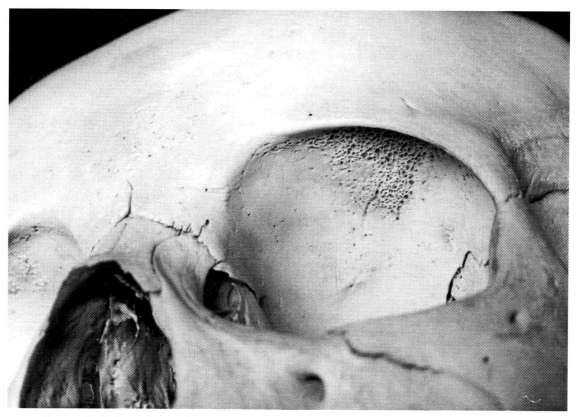

Figure 13. Cribra orbitalia (Ursa orbitale. Møller-Christensen and Sandison 1963). (NMNH)

1977; Fairgrieve and Molto 2000; Glen-Haduch et al. 1997; Guidotti 1984; Hengen 1971; Lanzkowsky 1968; Mittler and Van Gerven 1994; Robb et al. 2001; Schultz 2001; Steinbock 1976:239–248; Zaino and Zaino 1975. (See Porotic hyperostosis.)

The septum will deviate to either side of the midline of the nasal aperture and usually presents no clinical problem except when the deviation is so severe that it blocks the nasal passage (**Figure 14**). With the fact of normal asymmetry of the human face, some level of deviation in the nasal septum will be observed in all individuals. Collett et al. 2001; Guyuron et al. 1999; Kayalioglu et al. 2000.

Typically, erosion of the nasal spine is one of the early bony changes of the nasal area, followed by erosion of the inferolateral border of the aperture (rhinomaxillary change; Manchester 1989) (**Figure 15**). In archeological settings, the nasal spine often does not survive normal taphonomic erosion, and thus erosion in this area and the absence of the nasal spine should be carefully studied to determine whether it is pathological or pseudopathological. Rare finding in most populations and uncommon to common in Europe (depending on the time period).

Erosion of the nasal area is one symptom of syphilis, *facies leprosa,*

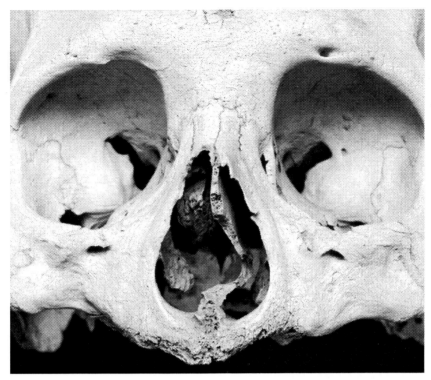

Figure 14. Deviated nasal septum. (NMNH 264502)

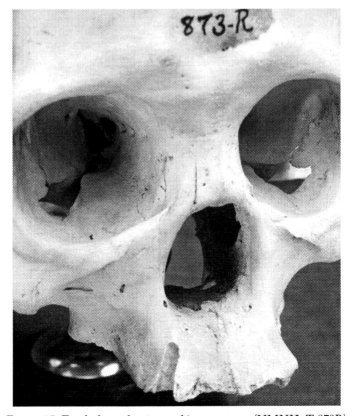

Figure 15. Eroded nasal spine and/or aperture. (NMNH–T 873R)

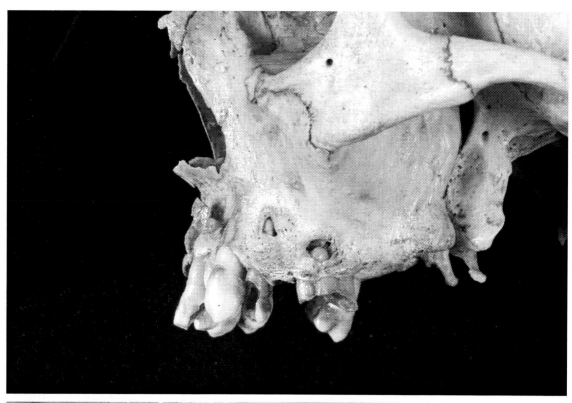

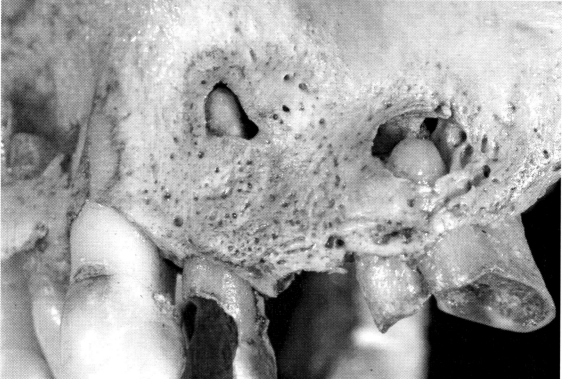

Figure 16a & b. Two periapical abscesses– inflammation at the apex of a tooth root with sinus formation (large to small pocket) and penetration of the maxilla or mandible. (NMNH–T 176R)

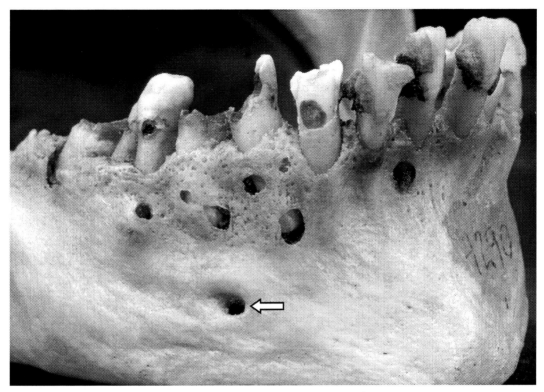

Figure 17. Multiple (at least 7) periapical abscesses in the mandible (arrow points to the normal mental foramen). (NMNH–T 1290)

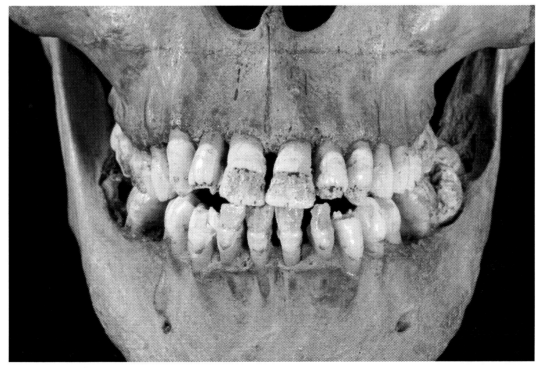

Figure 18. Linear enamel hypoplasia (hypoplastic lines, enamel dysplasia)–horizontal grooves resulting from a disturbance in the development of the enamel. (NMNH–T 707)

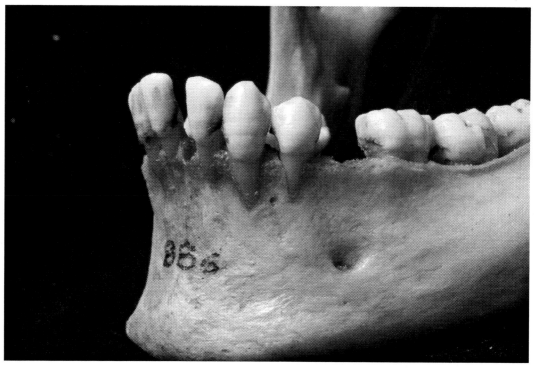

Figure 19. Multiple slight linear enamal hypoplasias on the dental enamel. (NMNH–T 886)

tuberculosis, and leishmaniasis among other conditions (Anderson 1989; Ell 1988; Manchester 1989; Møller-Christensen 1961, 1974, 1978; Reichart 1976; Queneau et al. 1982.). Extensive erosion is a rare finding.

The margins of a periapical abscess (hole) will exhibit some periosteal reaction (pitting), and a pocket may be visible at the root apex (**Figures 16, 17**). The margins of a healed abscess will be smooth, rounded and of similar texture as the surrounding bone.

Hypoplastic lines appear as subtle to deep grooves encircling the tooth crowns at the same level (**Figures 18, 19**). Enamel hypoplasia can be caused by many factors including periapical inflammation or trauma to a deciduous tooth, fever, disease, nutritional deficiencies (especially A and D), endocrine dysfunction, and generalized infection during odontogenesis (Robinson and Miller 1983). A second form of hypoplasia, not shown, is represented by pits of various sizes in the enamel. Common finding in many populations (e.g., American Indian). Blakey et al. 1994; Corruccini et al. 1985; Cucina 2002; Cucina and Iscan 1997; Goodman and Armelagos 1985, 1988; Goodman and Rose 1990, 1991; Goodman and Song 1999; Goodman et al. 1980, 1987; Hillson and Bond 1997; Hutchinson and Larsen 1988; Lawson 1967; Pindborg 1970; Reid and Dean 2000; Robb et al. 2001; Sarnat and Schour 1941.

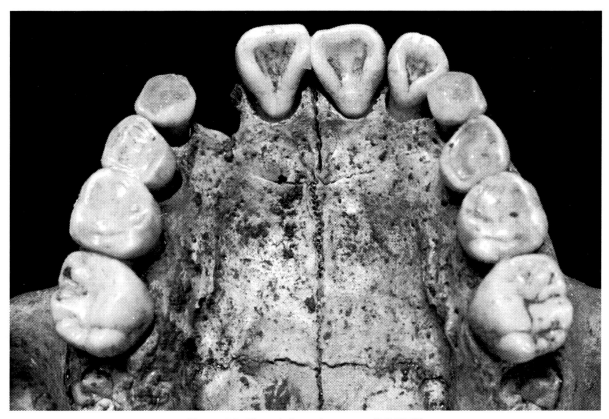

Figure 20. Shovel-shaped incisors. (NMNH)

Shovel-shaped incisors, so called because the lingual surface is shaped like a shovel (**Figure 20**). This is a dental trait found commonly in Asian groups (ranging from above 60% to 100%), while Whites and Blacks may exhibit the trait but to a lesser degree (generally below 5%). Dahlberg, 1951; Devoto 1971; Devoto and Arias 1967; Hinkes 1990; Hrdlicka, 1920; Hsu et al. 1997; Saini et al. 1990; Saunders and Mayhall 1982; Sharma 1983; Tsai and King 1998; Ubelaker 1999.

Look for asymmetry of the nasal bones including depressions, small adhering bone fragments, and fractures with evidence of healing (**Figure 21**). Common finding in most populations.

Some ethnic groups intentionally remove (extract or knock out; ablation) one or more teeth for aesthetic and cultural reasons (Chimenos-Kustner et al. 2003; Fitton 1993; Gould et al. 1984; Handler et al. 1982; Nelsen et al. 2001; Stewart and Groome 1968; Sweet et al. 1963) (**Figure 22**). In questionable cases, histological (Schultz 2001) and radiological analysis can help distinguish resorption due to cellular activity verses erosion due to acidic soil or an abrasive object.

Cleft palate (**Figures 23, 24**), one of the most frequently encountered congenital malformations, is a common structural defect that

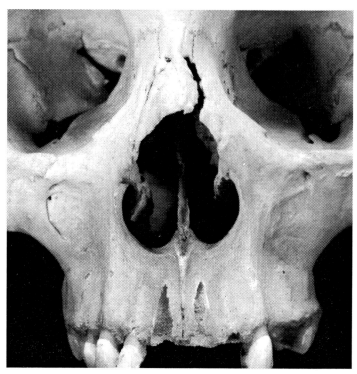

Figure 21. Fractured nasal bone(s). (NMNH–T)

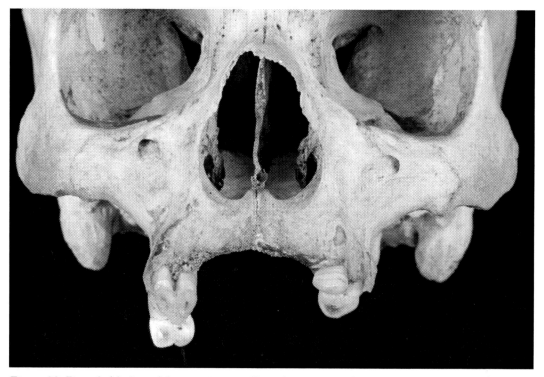

Figure 22. Dental ablation of the anterior maxillary dentition in a historic American Black. (NMNH 387866)

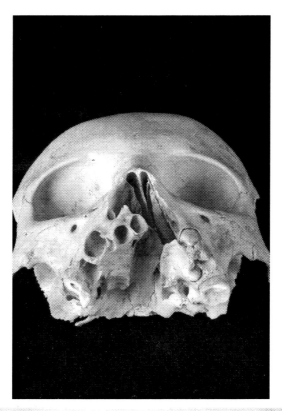

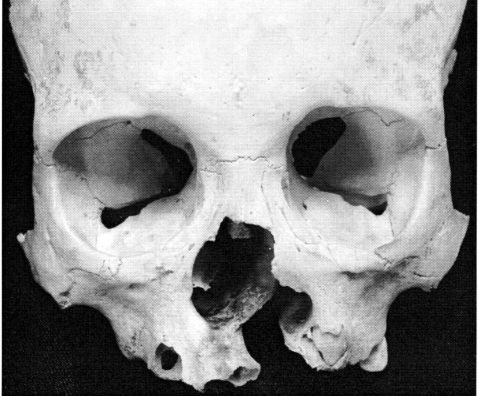

Figure 23a & b. Cleft palate (uranoschisis) (unilateral expression). (NMNH 293252)

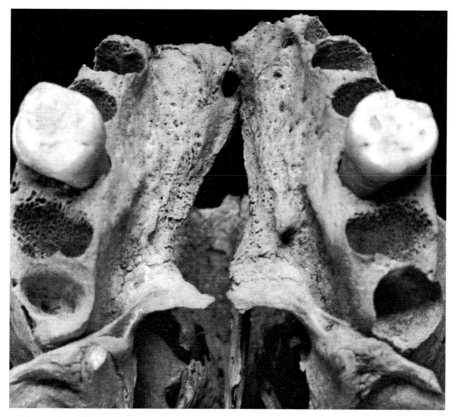

Figure 24. Cleft palate in the mid and posterior hard palate. (NMNH 316482)

results from faulty development (failure to fuse, interact or merge) of the soft (lip and soft palate) and hard tissues (maxilla/palate) of the oral cavity between weeks eight and twelve of pregnancy. Cleft lip with cleft palate occurs in about 1:1000 Caucasian births (O'Rahilly and Muller 1996), 1/1500 in Japanese and about 1/5000 in American Blacks (Bergsma 1978). Females seem to be more prevalent to cleft palate than males (Fishbien 1963). The expression may be unilateral or bilateral, and appears as a separation, however small, between the tooth sockets, or a separation in the posterior portion of the hard palate (O'Rahilly and Muller 1996). Cleft lip without cleft palate occurs in about 1:2500 births (60% to 80% affected infants are male; O'Rahilly and Muller 1996) and can result from a variety of genetic and environmental factors that alter palatal growth and formation in the developing embryo (Bender 2000; Coleman and Sykes 2001; Johnston and Millicovsky 1985). While cleft lip and cleft palate often occur together, they differ in their distribution regarding sex, familial occurrence, race, and geography (Larsen 2001). (See Tessier (1976) for classification of clefts, and Turvey et al. (1996) and Cohen and MacLean (2000) for photographic examples of facial clefts and cran-

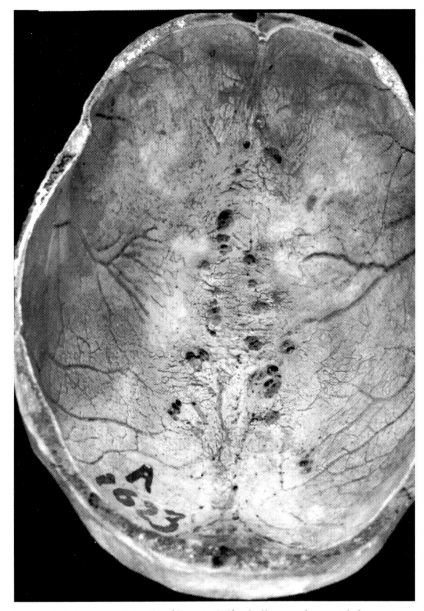

Figure 25. Lacunae laterales (Grant 1972)–shallow endocranial depressions (possibly raised "mound" ectocranially) located on either side of the sagittal suture. The pits are aracchnoid granulations. (AFIP A1623)

iosynostoses.) Barnes 1994:180–192 (see Fig. 4.27 for drawing of the range of forms of clefting); Goodman and Gorlin 1977; Ortner 2003:456–459.

Although these depressions vary in size, most are large, smooth bordered, and serve to house large clusters of arachnoid granulations (Pacchionian granulations, Pacchionian bodies) (**Figure 25**). Pacchionian pits, on the other hand, are small, sometimes clustered endocranial pits with sharply defined margins caused by erosion by small clusters of arachnoid granulations. Lacunae laterales are always located post-bregmatic (in the anterior parietals only) while Pacchionian pits may be found in the parietals (possible within the lacunae laterals) and frontal. Fox et al. 1996.

There are a number of theories as to the etiology for the erosion of the inner vault of the skull due to arachnoid granulations. It is known, however, that these granulations primarily serve to filter and return cerebrospinal fluid. In some cases the lacunae laterales may become eroded and result in localized protrusion and perforation of the outer vault by the cauliflower-shaped and ossified arachnoid granulations. Both depressions (lacunae) and pits are extremely common in all populations and increase in number and depth with age. Perforations of the outer vault that are not due to postmortem erosion are rare. (See Pacchionian pits.) Pits resulting from erosion of the inner table of the vault due to enlargement and ossification of arachnoid granulations that serve to filter cerebrospinal fluid. In young individuals arachnoid granulations are villous and small. During old age the granulations enlarge, become cauliflower-shaped, and erode the cranial vault resulting in varying sized pits. Pacchionian pits appear as relatively small (2 mm) to large (5 mm) pits with sharply defined margins that are mostly confined to the parietals. The depth and frequency of these lesions increase with age and, possibly, disease. Common finding in all populations.

Vascular/arterial grooves–shallow grooves, usually symmetrically situated in both parietals, lateral and posterior to the parietal foramina.

Venous lakes–normal vascular structures situated at the end (commonly) or along the path of one or more arterial grooves. Grossly, these irregularly shaped depressions in the inner table often resemble cauliflower and as radiolucencies radiographically. They often penetrate through the inner table into the diploe and sometimes perforate the outer table.

This area may either be flat (typically in females) or developed and projecting inferiorly in some males (**Figure 26**). Heavily developed nuchal crests may have a shelf-like ridge and an inferiorly oriented "spike/spine" or Inion hook of bone in response to muscular activity and bone stimulation and is found more commonly in males (63%) and 4.2 percent in females (Gulekon and Turgut 2003). Common finding.

This feature is often observed on crania which have had fronto-occipital or occipital cranial deformation (**Figure 27**). (Personal communication, K. Murray 1989) Stewart 1976.

Craniosynostosis is a term commonly used to refer to any suture that fuses at too early an age (**Figure 28**). While most authors use the

Figure 26. Nuchal or Inion spike. (NMNH 387865)

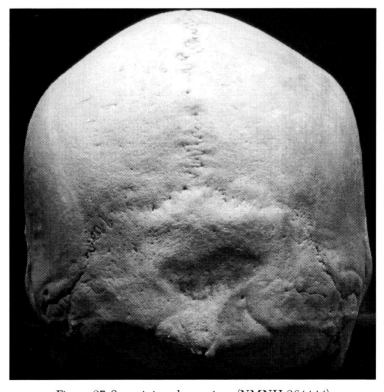

Figure 27. Suprainion depression. (NMNH 264444)

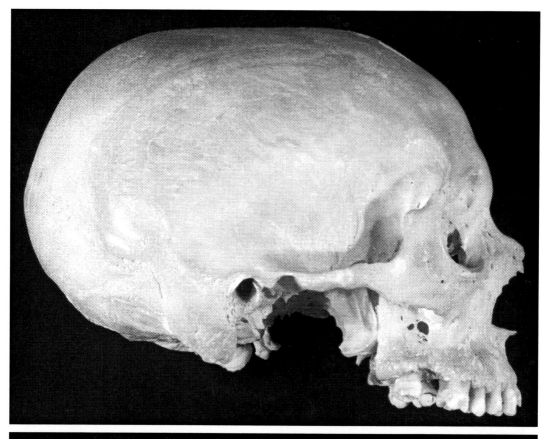

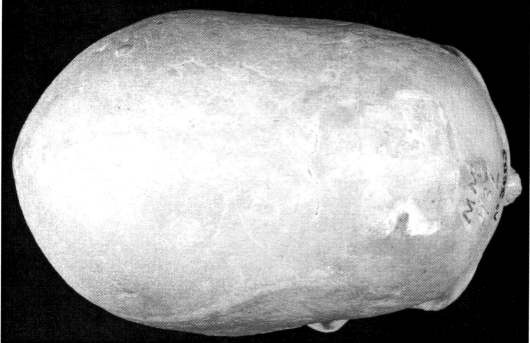

Figure 28a & b. Scaphocephaly (craniosynostosis, craniostenosis, premature suture closure). (AFIP MM737/ A3603) (See also Figure 40a–c)

terms craniosynostosis and craniostenosis interchangeably, others (Prokopec et al. 1984) use the latter term only when the cranial capacity is diminished. When the sagittal suture prematurely fuses, the skull continues to grow and results in a long-headed individual (scaphocephaly or hyperdolicocephaly). Premature closure of the coronal suture results in a high (pointed) skull known as steeple skull, tower skull, or oxycephaly (Silverman and Kuhn 1993). Craniosynostosis does not refer to artificial cranial deformation from such practices as cradle boarding or head binding (Dingwall 1931; Ozbek 2001; Stewart 1941). Craniosynostosis is an uncommon finding in most skeletal populations and may be congenital, hereditary, or the result of metabolic disturbances. Craniostenosis is more common in males (Cecil and Loeb 1951) and has been associated with Crouzon syndrome (Gorlin et al. 1976). See Barnes 1994; 152–175 for an in-depth discussion of developmental problems causing agenesis and stenosis with reference to paleopathological cases and photographic illustrations. Aufderheide and Rodriguez-Martin 1998:52–55; Barnes 1994:152–157; Chopra, 1957; Cohen 1986; David et al. 1982; Moore 1982; Prokopec et al. 1984; Stewart 1975, 1982; Turvey et al. 1996; Webb 1995:251 (for severe deformation from scaphocephaly in an Aboriginal Australian).

Scaphocephaly is one of the most common forms of craniosynostosis. (The authors have come across this form in seven crania, five of which were Black). Some of the traits associated with this condition consist of a bulbous, projecting frontal bone and low-set eye orbits in relation to the frontal bone. Bilateral fusion of only the coronal suture is sometimes referred to as brachycephaly in the old literature. Unilateral fusion of the coronal is referred to as plagiocephaly (**Figure 54a & b**). Barnes 1994:152–157; Cohen 1986; Cohen and MacLean 2000; David et al. 1982; Moore 1982; Prokopec 1984; Tod and Yelland 1971; Turvey et al. 1996.

Inflammation and subsequent infection of the middle ear (otitis media) may result in perforation and resorption of the mastoid process or other portions of the temporal bone (**Figure 29**). The middle ear is an air-filled space within the petrous portion of the temporal bone lying immediately behind the tympanic membrane (eardrum). Otitis media is very common in infants beyond the neonatal period (after 28 days of age) and shows a decline in incidence after the first year of life (Bluestone and Klein 1988). Skeletal involvement of the mastoid, however, is uncommon to rare in most skeletal samples. Care must be taken not to mistake the normal fissures and squamo-mastoid sutures (see Hauser and DeStefano 1989:196–207 for normal variants) in the outer surface to the mastoid for a pathological condition. Also be careful not to confuse the numerous air cells of the normal mastoid for disease.

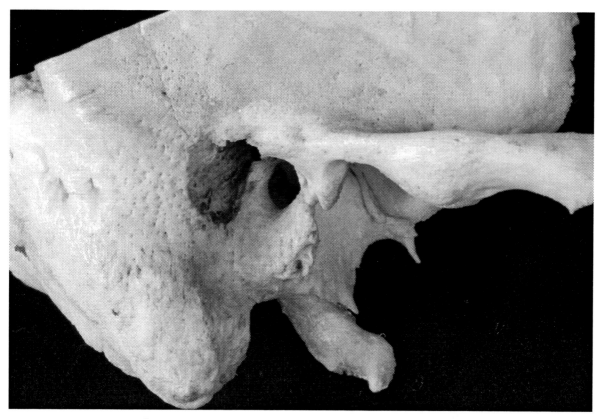

Figure 29. Mastoiditis resulting in otitis media. (NMNH–H)

When mastoiditis is suspected, use radiographs to look for sclerosis and pocket formation. Dugdale et al. 1982; Edwards 1988; Gregg and Gregg 1987; Paparella et al. 1980; Schultz 1979; Teele et al. 1984.

This tumor (**Figure 30**) is easily visible as a rounded lump ranging in size from small (barely visible) to large, virtually filling the opening. Rarely the external auditory meatus may be absent at birth; no external canal will be seen in the temporal bone (the authors have encountered two such cases). Hrdlicka (1935) identifies exostoses to be seen in much higher frequencies in males than in females and that they appear to increase in size with age. Auditory exostoses are uncommon to common findings found in higher frequencies in swimmers and divers, especially in cold water (Arnay-de-la-Rosa et al. 2001; Filipo et al. 1982; Kroon et al. 2002) and have been attributed to both genetic (Hanihara and Ishida 2001a) to behavioral traits (Kennedy 1986). Otolaryngologist DiBartolomeo (1979) found 70 cases of auditory exostosis in 11,000 patients over a period of 10 years (6.4 per 1,000 patients) in a coastal region. He reported that "irritation nodules" in the external auditory canal are painless until the tenth year of aquatic exposure, at which time obstruction of hearing may occur. For other case reports see ref-

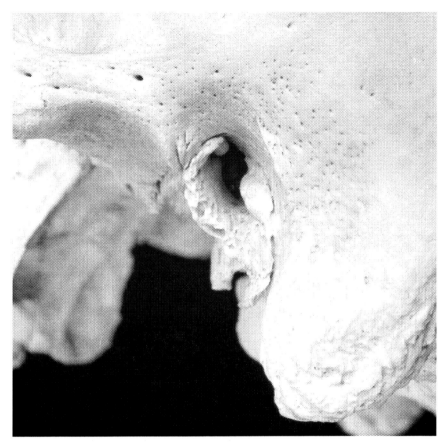

Figure 30. Auditory exostosis (torus)–benign bone tumor(s) of the ear canal. (NMNH)

erences by Pulec and Deguine (2000; 2001) and Deguine and Pulac (2001). Wong et al. (1999) found a relationship between the length of time spent surfing and the severity, location, and prevalence of auditory exostoses. Fenton et al. 1996; Graham 1979; Gregg and Bass 1970; Gregg and Gregg 1987; Hanihara and Ishida 2001; Hrdlicka 1935; Hutchinson et al. 1997; Kennedy 1986; Longridge 2002; Pulec and Deguine 2001; Steinbock 1976:329–333.

To identify the cause of perforations in the cranial vault, look for new bone spicules encircling the lesion (both endo- and ectocranially), increased vascularity (tiny, smooth-bordered pits), and remodeling (healing or filling in) along the margin of the lesion (**Figure 31**). Some of the more common diseases that produce these lesions are metastatic carcinoma, tuberculosis, multiple myeloma, eosinophilic granuloma, and fungal infections. All of these types of lytic lesions cause destruction of bone that can be mistaken for postmortem damage (see **Figure 53**). Seek the assistance of an experienced researcher before drawing any conclusions. Apley and Solomon 1988; Grupe 1988; Mays et al. 1996;

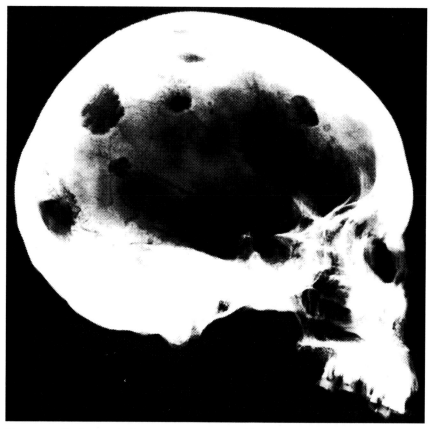

Figure 31. Radiograph illustrating perforation(s) due to disease of the cranium—diagnosing diseases based solely on examination of lytic lesions in the skull is extremely difficult.

Olmsted 1981; Ortner 2003; Sefcakova et al. 2001 (for an interesting case and presentation of metastatic carcinoma in ancient remains); Silverman and Kuhn 1993; Steinbock 1976; Webb 1995.

Caution must be exercised when making a differential diagnosis based solely on examination of the skull. Carefully describe the appearance and distribution of the lesions throughout the skeleton. As an example, in comparison to multiple myeloma, histiocytosis X of the skull usually appears as a single lytic lesion with a central island of bone/sequestration. There are a number of other diseases that produce similar lesions including histiocytosis X (more correctly called Langerhans cell histiocytosis (Ladisch and Jaffe 1989; Taylor and Resnick 2000)), multiple myeloma, fungal infections, and tuberculosis. The distribution of the lesions in the skeleton must be considered as well as their radiographic appearance. Sex may also affect the expression; in males, prostate cancer causes osteoblastic response at the metastitsis, while in females the metastitsis will have both osteoblastic

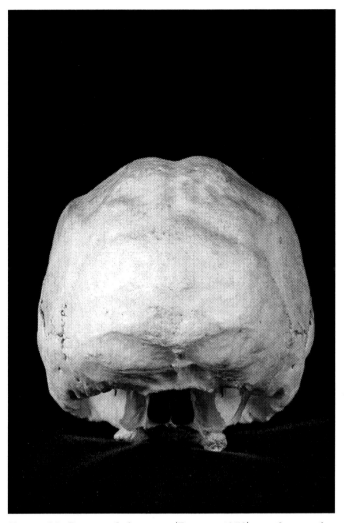

Figure 32. Biparietal thinning (Epstein 1953); senile atrophy (Wilson 1944)–symmetrical depressed areas in the posterior outer table of the parietals. (NMNH–T 878)

and osteoclastic responses from breast cancer. Age may determine the cancer. For example, if a child of less than two or three years of age exhibits these lesions, the acute phase of histiocytosis X (Letterer-Siwe) might be suspected while in adults, Hand-Schuller-Christian disease (the chronic stage of histiocytosis X) might be the proper diagnostic term (Lichenstein 1970; Taylor and Resnick 2000). Bone involvement in histiocytosis X is 80 percent (Ladisch and Jaffe 1989). Coley, 1960; David et al. 1989; Jaffe 1975; Lichenstein 1953; Moseley, 1963; Olmsted 1981; Ortner and Putschar 1985; Steinbock 1976; Wroble and Weinstein 1988.

The parietal bones at the lateral bossing and superior posterior regions will be extremely thin, fragile, and translucent (**Figure 32**).

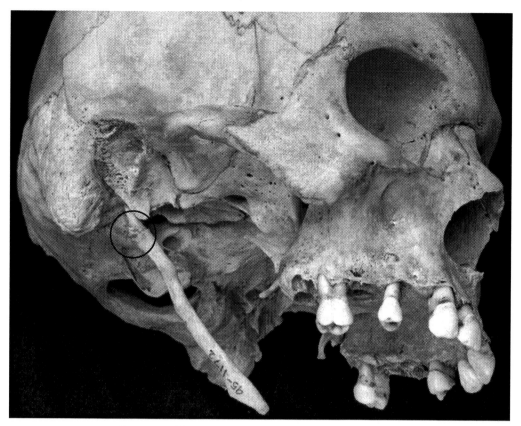

Figure 33. Elongated styloid process. (AFIP 95–1172)

The process of atrophy begins in the outer table (Wilson 1944). This fairly rare condition has unknown etiology but appears to be age related (Grainger et al. 2001) but alternative theories that this is the result of endocrine imbalance has also been postulated. The youngest affected individual known to the authors is 40 years of age. An interesting note is that the thinning usually avoids the parietal foramina leaving approximately 1 cm of undisturbed bone encircling these foramina. Barnes 1994:146–148; Brothwell 1967; Hauser and DeStefano 1998:83–84; Lodge 1967; Ortner 2003:415.

Elongation of the stylohyoid process is due to ossification or calcification (mineralization) of the stylohyoid ligament (**Figure 33**). The normal length of the styloid process is 2.5–3.0 centimeters in length. Note that in the present case, the calcification has been interrupted and a pseudarthrosis formed at the junction (Type II). Type I defects consist of uninterrupted styloids. Type III ("segmental") defects consist of elongated styloids with multiple interrupted segments of ossified stylohyoid ligaments. Even in severe cases of stylohyoid ligament ossification, more than 50 percent of individuals are asymptomatic (White

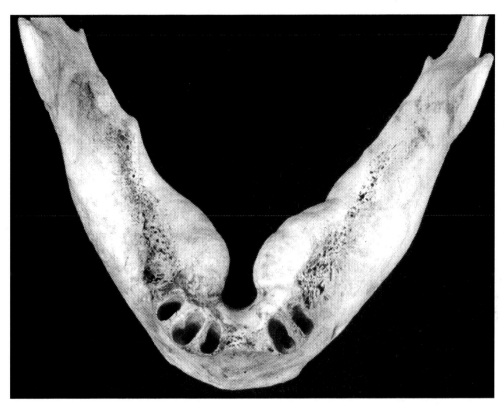

Figure 34. Torus mandibularis. (NMNH)·

and Pharoah 2000). Others, however, may be accompanied with pain, impingement of the carotid artery, sore throat, the sensation of a "foreign body" lodged in the throat, vertigo, and other symptoms and are referred to as Eagle's syndrome (Chi and Harkness 1999; Eagle 1948; Kay et al. 2001; Keur et al. 1986; Langlais et al. 1986; Monsour and Young 1986; Prasad et al. 2002; Restrepo et al. 2002; Sivers and Johnson 1985; Takada et al. 2003; Thot et al. 2000). Unexpectedly, the length of the elongated styloid and severity of pain does not appear to be correlated. Symptoms of Eagle's syndrome show no sex predilection and appear after age 30. One study revealed calcificatioin of the stylohyoid ligament in 40 percent of children with a mean age of 11 years (Camarda et al. 1989). While it is not possible to reliably ascertain whether an individual suffered Eagle's syndrome from skeletonized remains, the styloid should be described and measured. DeChazal 1946; Douglas 1952; Feldman 2003; Ferrario et al. 1990; Omnell et al. 1998; Patni et al. 1986. Styloid processes, especially elongated forms, are prone to fracture and healing. These fractures are difficult to detect due to the irregular shape and appearance of the process (Haidar and Kalamchi 1980).

Rounded, usually symmetrical bony growths (hyperostoses or

hamartomas) may be found along the superior lingual border of the mandible below the premolars, but may extend further along the mandibular corpus (**Figure 34**). The tori arise from the cortical plate and are developmental anomalies that may extend in the form of two bulging ridges that nearly touch one another behind the incisors. If the tori are quite large, restriction of the tongue may occur. Chronic gum disease (gingivitis), however, may also stimulate similar, albeit, less pronounced growths. Jainkittivong and Langlais (2000), in a study of 960 Thais, found that exostoses were more common in men than in women, suggesting an interplay of multifactorial genetic and environmental factors. The presence of tori increase in frequency with age. Tori are most commonly found in Eskimo populations, but can range in frequency depending on the population from nine to 66 percent depending on ethnicity (Seah 1995; Hauser and DeStefano 1989:184–5). Shah et al. (1992) reported that tori in a sample of 1,000 patients in India were rarely present before 10 years of age. Eggen and Natvig (1986) found a positive correlation between torus mandibularis and the number of functioning teeth in a sample of 2,010 dental patients over 10 years of age. Antoniades et al. 1998; Hauser and DeStefano 1989:182–185; Pynn et al. 1995; Seah 1995.

Stafne's defect vary in size and shape as: (1) a circular or oval, smooth-walled concavity varying in size from 1–2 mm to more than 1 cm in diameter and located in the lingual surface of the mandible inferior to the mylohyoid line; or (2) a shallow, circular, or oval and roughened defect less than 1 mm deep in the location noted above (**Figures 35–36**). Radiographically, the defect has a sclerotic border. The location and radiographic appearance of these lesions are highly suggestive, although not pathognomonic (diagnostic), of Stafne's defects. Although the etiology of Stafne's defect is unknown, nearly all have proven to be benign defects, not tumors or cysts (Thawley et al. 1987), containing normal submandibular salivary gland tissue. The osseous defects result from pressure erosion of the mandible by the submandibular salivary gland or duct.

Preliminary examination (RWM) of over 5000 dry-bone mandibles from historic and prehistoric sites revealed 91 individuals with defects, of which 81 were males (17 individuals from the same site in Alaska). Numerous researchers have confirmed the predominance of this trait in adult males; most individuals are in their forties and fifties when the defects are first detected. Stafne's defects are developmental rather than congenital or traumatic in origin and fit the expected frequency of an X-linked recessive trait (Mann 2001). The youngest known individual with a Stafne's defect was an 11-year-old boy from Sweden (Hansson 1980). Stafne's defects are usually unilateral. Uncommon

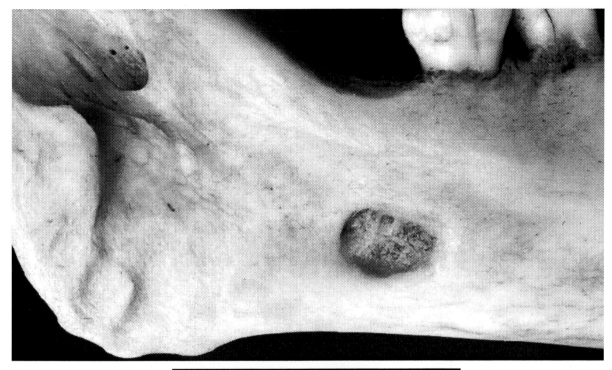

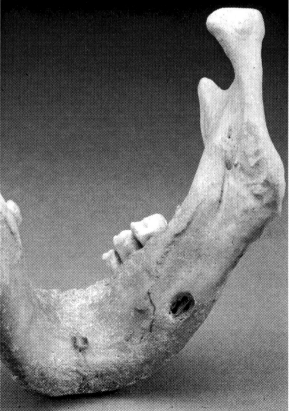

Figures 35 & 36. Stafne's defect (static bone defect, posterior lingual depressions, lingual cortical mandibular defects, salivary gland defect, latent bone cyst). Figure 36 photographed by Chip Clark.

finding in most skeletal samples. Correll et al. 1980; Finnegan and Marcsik 1980; Gorab et al. 1986; Harvey and Noble 1968; Mann 2001; Pfeiffer 1985; Shields and Mann 1996; Stafne 1942; Tolman and Stafne 1967; Uemura et al. 1976.

Although a true Inca bone is where the mendosa suture has been retained and extends from asterion to asterion, Inca bones are identified by most researchers as any large accessory bone(s) in the lambdoidal region of the occipital bone (**Figures 37, 38, 39**). The common presence of Inca bones in most world populations, including Sub-Saharan Africans, indicate that this trait is not uniquely Asian (Hanihara and Ishida 2001). Inca bones are found to be more prevalent in males (Berry 1975, Hauser and DeStefano 1989:103). Barnes 1994:140; El-Najjar and Dawson 1977; Hauser and DeStefano 1989:99–103 (see Plate X and Fig. 12 for illustrations of different degrees of expression); Matsumura et al.

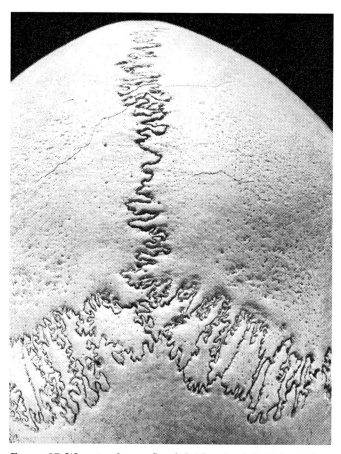

Figure 37. Wormian bones (lambdoid or lambdoidal ossicles, sutural bones)–small to large bones that may persist as separate ossicles or unite with the parietal and occipital bones. Hauser and DeStefano 1989:84–103 (Fig. 15 for range and location of ossicles in the lambdoidal suture).

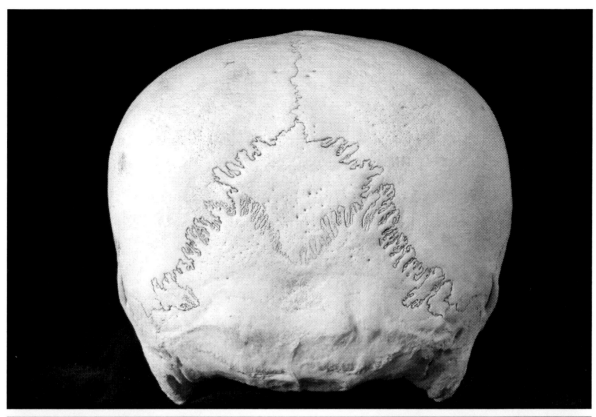

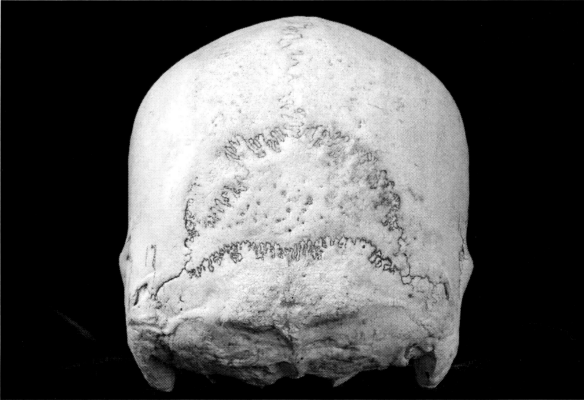

Figures 38 & 39. Inca bone (os incae). (NMNH 264539, and 264541, respectively)

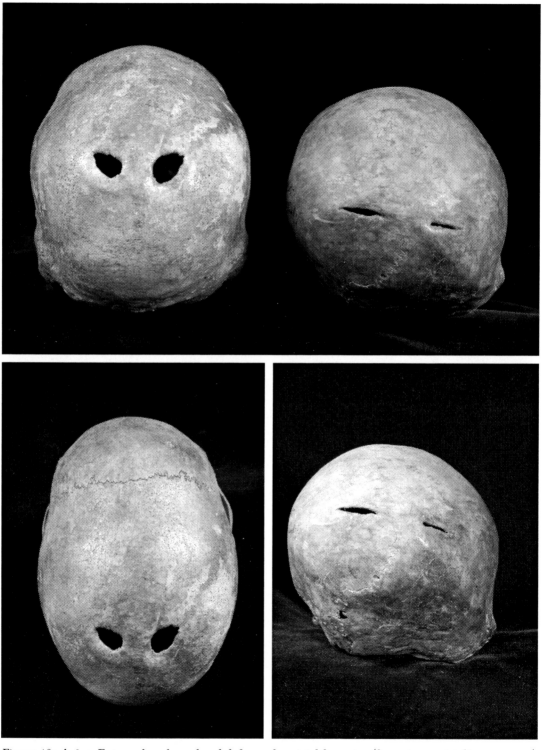

Figure 40a, b & c. Extremely enlarged and deformed parietal foramina (foramina parietalia permagna) (NMNH 276981, left and 276982, right). Figure 40b is a superior view of NMNH276981 and Figure 40c a posterior view of NMNH 276982. Notice sagittal stenosis in the adult and in the subadult additional stenosis of the right lambdoidal suture with skewing of the right temporal bone. See discussions on parietal foramina by Goldsmith 1922; Hoffman 1976.

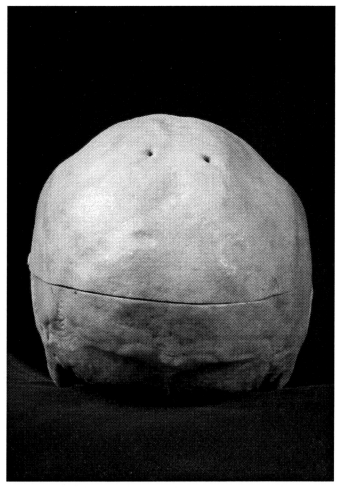

Figure 41. Enlarged paritetal foramina in an older adult with parietal thinning. (NMNH–T 1463)

1993; O"Rahilly and Twohig 1952).

These defects are usually bilateral bone defects in the posterior portion of the parietal bones (**Figures 40, 41**). These defects, caused by incomplete or faulty ossification, range in size and shape from small to large circular or oval perforations measuring several centimeters in diameter, to slits. Although usually benign and asymptomatic, enlarged parietal foramina have been found in association with other malformations. Greatly enlarged foramina (foramina parietalia permagna) reflect a hereditary condition, which may be transmitted in family members as an autosomal dominant trait (Dzurek et al. 1956; Fein and Brinker 1972; Goldsmith 1922; Hoffman 1976; Sjøvold 1984). This variant is found more often in males. While small perforations are found in 60 percent to 70 percent of normal skulls, variants larger than 5mm are less common with a prevalence of 1:15,000 to 1:25,000 (Kortesis et al. 2003).

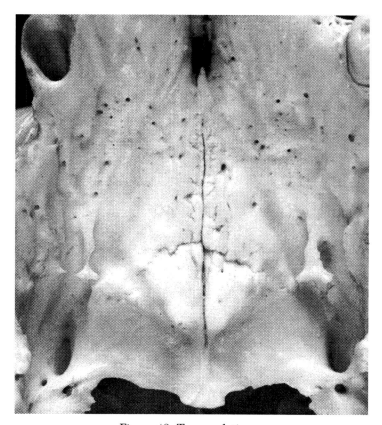

Figure 42. Torus palatinus.

Barnes 1994:143–146; Dzurek et al. 1956; Hauser and DeStefano 1989:78–82; Hollender 1967; Kutilek et al. 1997; Lodge 1967; Mann 1990; Murphy and Gooding 1970; O'Rahilly and Twohig 1952; Pang 1982; Pepper and Pendergrass 1936; Rasore-Quartino et al. 1985; Stallworthy 1932; Stibbe 1929; Symmers 1895; Wuyts et al. 2000; Zabeck 1987.

A raised plateau (**Figure 42**) of bone varying in size along the midline of the palate (not to be confused with the bony build-up along the anterior median palatine suture). The frequency and severity of this trait varies. Chohayeb and Volpe (2001), for example, studied the presence of torus palatinus in five ethnic groups (448 women) residing in Washington, DC and found the highest frequencies in African Americans, followed by Caucasians, Hispanics, Asians and Native Americans. They found no statistically significant relationship between age and the presence of torus palatinus. Antoniades et al. 1998; Eggen and Natvig 1994; Gorsky et al. 1998; Hauser and DeStefano 1989:174–179 (Plate XXVI illustrates different degrees of expression); Jainkittivong and Langlais 2000; MacInnis et al. 1998; Vidic 1966.

Perforation of the tympanic plate of the temporal bone (**Figure 43**).

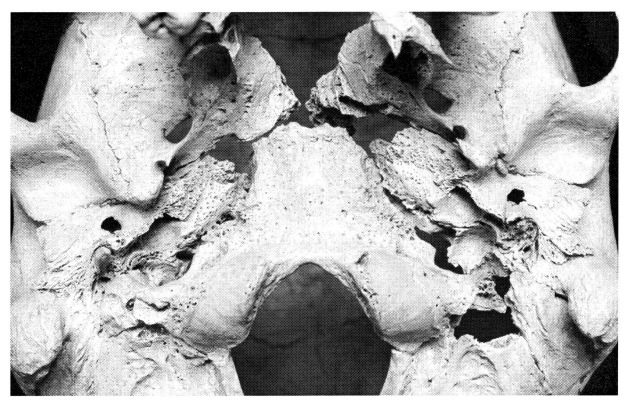

Figure 43. Bilateral tympanic dehiscences (foramen of Huschke). (NMNH)

The dehiscence appears as an irregular circular hole, posterior to the temporal fossa (TMJ). This is a developmental aberration with failure of closure of the foramen of Huschke that is seen in all young children that only occasionally persists beyond five years of age (Anand et al. 2000; Berry and Berry 1967; Wang et al. 1991). Wang et al. (1991) found the foramen of Huschke in 7 percent of adult skulls in a sample of 377 dry skulls. There appears to be no sex specificity, seen in higher frequency in males by some investigators (Krogman 1932; Laughlin and Jorgenson 1956) while higher frequencies were observed by others (Corruccini 1974; Dodo 1972; Molto 1983). Uncommon to common finding. Gerszten et al. 1998; Hauser and DeStefano 1989:143–147; Herzog and Fiese 1989; Sharma and Dawkins 1984.

Premature closure of the cranial sutures (e.g., lambdoidal, temporal or coronal) affects the overall structure and morphology of the cranium and can also result in an asymmetrical cranial base (**Figure 44**). Also referred to as torticollis or Rye Neck syndrome (Canale et al. 1982; Klepinger and Heidingsfelder 1996), this asymmetric postion of the cranial base and occipital condyles affects the postion of the body and thus changes in the spine may also be observed. Uncommon fining in most populations but frequently seen in Eskimos (experience of

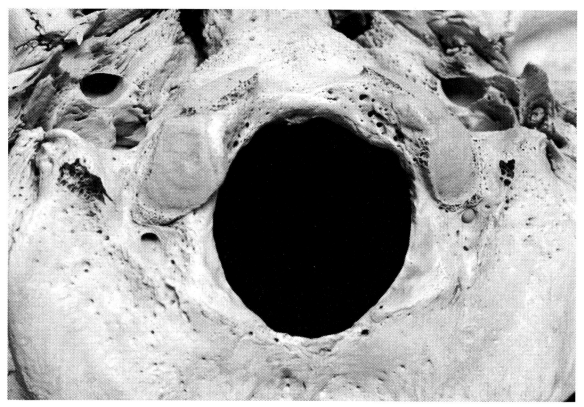

Figure 44. "Tilt" head-twisted and asymmetrical occipital condyles and foramen magnum referring to a condition initially identified in Hawaiian skulls by Snow (1974:61–67).

authors). Barnes 1994:136–137; Canale et al. 1982.

Normal temporal fossae typically have undulating surfaces but will not have the porosity or bony buildup (**Figures 45, 46**). Since the TMJ is a paired joint that cannot function alone, OA if present, is often symmetrical. Lesions are usually detectable earlier and are more severe in the temporal surface than in the mandibular condyle. If only one joint exhibits severe bone loss or bony remodeling look for signs of infection or trauma. OA of the TMJ is a common finding in most populations. Blackwood 1963; Markowitz and Gerry 1950; Ryan 1989.

The first indication of porosity (pits) can usually be seen in the middle of the fossa or on the articular eminence (the raised area just anterior to the fossa; when the head of the mandible anteriorly dislocates, it rides forward and up on this eminence). Usually there is a well-defined rim with a central depressed area representing early OA. Moderate OA are those cases where large areas of bone are missing due to erosion (flattened) or porosity. Severe OA of the temporal fossa presents when most of the fossa has been eroded, with or without the formation of osteophytes.

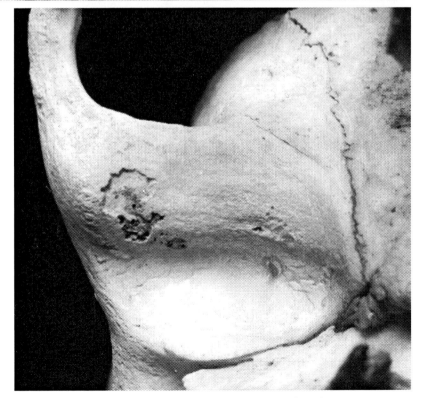

Figure 45a & b. Bilateral osteoarthritis of the temporomandibular joint (TMJ)–tiny to large pits and/or osteophytes on the articular surface and margin of the temporal fossa.

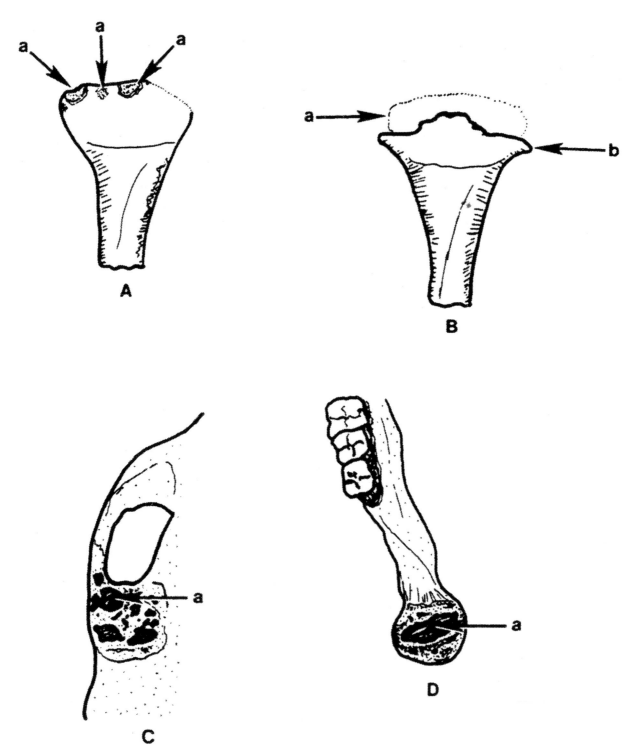

Figure 46. Forms of osteoarthritis/DJD of the mandibular condyle, fossa and eminence. Erosion and pitting (a) of the bone surface from slight pits (featured in A, to larger sclerotic edged erosion as featured in C and D. Complete loss of the joint surface from erosion and abrasion is found in more severe cases (featured in B and D). Osteophyes will form (b) along regions where reactive bone growth occurs.

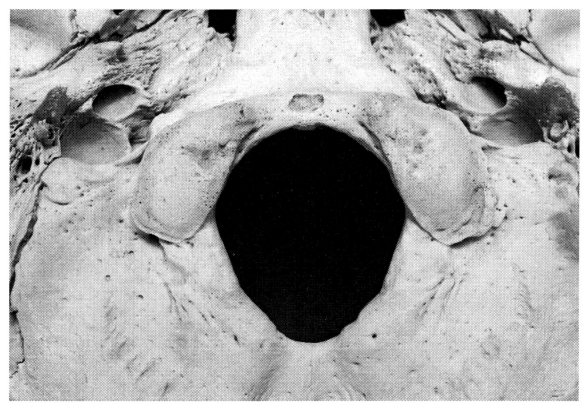

Figure 47. Precondylar defect at the anterior margin of the foramen magnum.

A small concavity in the precondylar area is one variant produced in the precondylar developmental formation (**Figure 47**). Caudal shifting in the development of the spine affects the morphology of this region, either producing a fossa or odontoid-like tubercle(s) (Barnes 1994:83–90; Hauser and DeStefano 1989:134–136). Incidence is low, generally not greater than 5 percent in a population. Barnes 1994:83–94.

This developmental trait should not be confused with an occasionally observed pseudarthrosis at this site produced by the dens when collapse or distortion of C1 allows the dens to ride on the precondylar region (see example in Barnes 1994:97, Fig 3.29) or by arthritic enlargement by osteophytic action of the dens (see **Fig. 68**) and the C1 articulation.

Look for irregularly shaped margins and extensions of bone, especially along the posterior margin of the condyles. **Figure 48** is close-up of left condyle from cranium pictured in **Figure 47**.

Irregular, undulating bony growths located on the inner surface of the vault, primarily in individuals heavily predilected to females (>95%) over 40 years of age (**Figure 49**). The etiology is unknown but Jaffe (1975) suggests the bone formation is a response to endocrine inbalance since it is seen in greater than 60 percent of postmenopausal women.

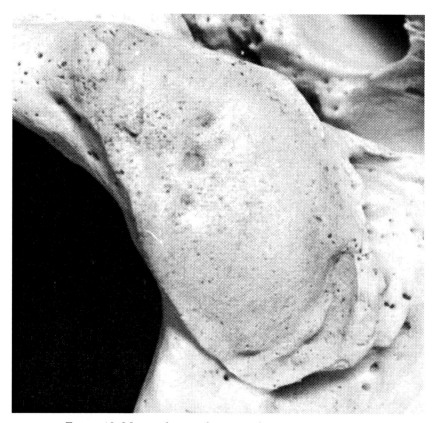

Figure 48. Marginal osteophytes on the occipital condyle.

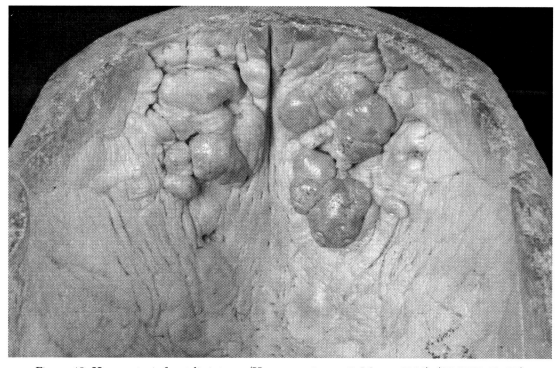

Figure 49. Hyperostosis frontalis interna (Hyperostosis cranii. Moore 1955). (NMNH–T 472)

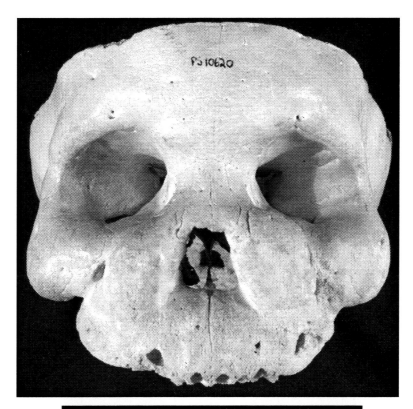

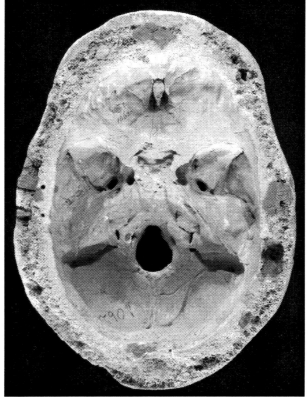

Figure 50a & b. Paget's disease (osteitis deformans, Paget 1877). (AFIP PS10620)

Uncommon finding. Most prevalent on the frontal bone, followed by the parietals. Gershon-Cohen et al. 1955; Perou 1964; Resnick 2002.

A chronic inflammatory condition that results in proliferation, thickening of the spongy bone in which this less organized deposition produces softened bone that may affect any or all bones in the skeleton (**Figure 50**). Mirra (1987) hypothesized that a unique slow-virus infection of osteoclasts causes Paget's. In the early stages of this disease the lesions are typically lytic (resorptive), originate in one focus of bone, and slowly spread until the entire bone is affected (Mirra 1987). Bone involvement generally occurs unilaterally at first. Late phases result in grossly enlarged, dense bones, especially noticeable in the skull and extremities. The disease seldom appears before the age of 40 years. In European populations, frequencies are found generally around 2 percent, but increase to approximately 10 percent as population age increases. Only about 5 percent of the individuals with the disease will express the manifestations of Paget's in the skeleton (Russell 1984). Both sexes are vulnerable to the disease, but males are more often affected (Jaffe 1975). Differential diagnosis should include metastatic carcinoma, hyperparathyroidism, leontiasis ossea and saber shin from syphilis (Zimmerman and Kelley 1982:58). Auferdeide and Rodriguez-Martin 1998:413–417; Barry 1969; Kaufmann et al. 1991; Lawrence 1970b; Mirra et al. 1995a; Mirra et al. 1995b; Moore et al. 1990, 1994; Nugent et al. 1984; Ortner 2003:435–443; Ortner and Putschar 1985; Singer 1977; Wells and Woodhouse 1975; Zimmerman and Kelly 1982:56–58.

Extensive destruction (cavitation) of the outer table and diploe, stellate scarring (radiating grooves), and nodules (mounds) in the outer vault caused by a variety of treponematoses (not to be confused with lesions caused by hemangioma or mengioma as in **Figure 52**). Note the smooth, raised, and rounded nodules (healed) as well as the depressed lesions that have eroded both the inner (rare) and outer tables. The late stage of syphilis (tertiary) (**Figure 51**) can affect any number of bones but is most frequently seen in the tibiae (thickening periostosis and "sabre shin" tibia) (see **Figures 191 & 192**) and skull. Erosive destruction of the skull may be evident in any area of the outer vault, maxilla, palate, malars, and nasal aperture. Aufderheide and Rodriguez-Martin 1998:157–164; Crane-Kramer 2000; Crosby 1969; Ortner 2003:280–283; Ortner and Putschar 1985; Stirland 1991a.

Carefully consider the processes which have caused erosion of the bone. A hand lens or magnifying light should be used to look for indications of taphonomic processes or osteolytic activity. As Schultz (2003) suggests histological evaluation should be employed to more accurately determine the cause of the lesion, or pseudo-lesion (**Figure

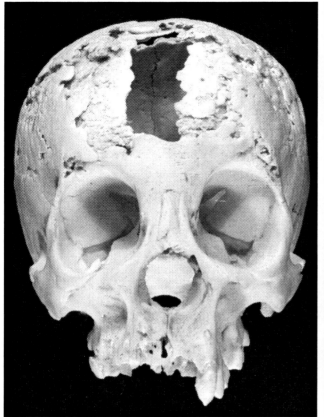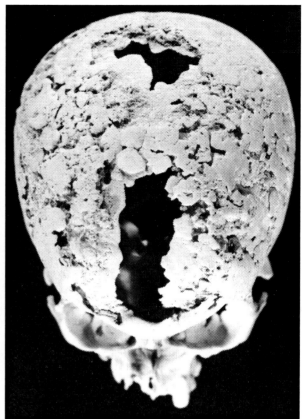

Figure 51a, b & c.
Tertiary syphilis of the skull. (AFIP)

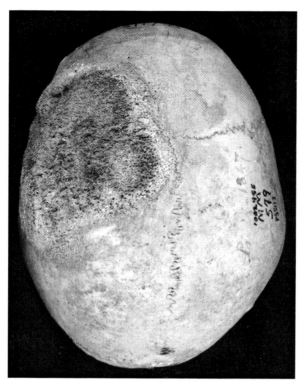

Figure 52. Exuberant hemangioma on the left frontal and parietal. (AFIP MM579)

Figure 53. Postmortem erosion of the left parietal and frontal bone resulting in pseudopathology resembling resorptive lesions similar to treponematosis. (NMNH 387867)

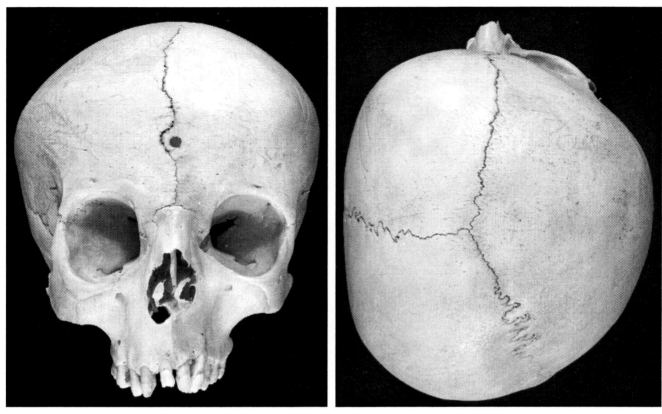

Figure 54a & b. Plagiocephaly (frontal and superior view). (AFIP)

53). Schultz 1997, 2003; Wells 1967.

Premature closure of one-half of the coronal suture (**Figure 54**) result-ing in an oblique (skewed) skull form (Aufderhede and Rodriguez-Martin 1998:53–4; Cohen and MacLean 2000; Perou 1964; Schendel et al. 1996; Turvey et al. 1996). Stenoses are more often seen in males (Cecil and Loeb 1951). Uncommon to rare finding (the authors have encoun-tered only one case in an American Indian). Barnes 1994:152–157.

This condition results in an unusually large, and often wide cranium and is in many cases incorrectly referred to as "hydrocephaly" when in fact the cause is not unknown (**Figures 55 & 56**). The cranial bones generally are thin and the regions of the fontanelles may have ossicles. Macrocephaly may be a response to a number of factors, possibly con-genital (since macrocephaly is more often identified in males (Zimmerman and Kelly 1982:26) but may be due to physical effects such as juvenile hydrocephaly ("water on the brain"). Macrocephaly is a rare finding that is often difficult to discern from proportionately large, yet normal skulls in children. Macrocephaly should be differen-tiated from acromegaly where the cranium is proportionally large as a

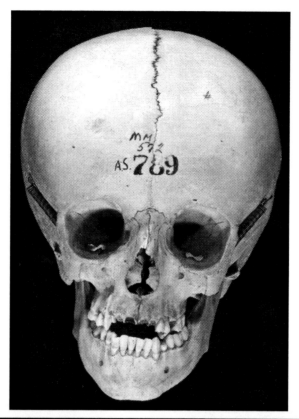

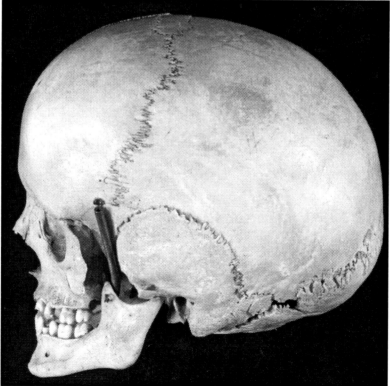

Figure 55a & b. Macrocephaly (frontal and left lateral view). (AFIP MM572/AS789)

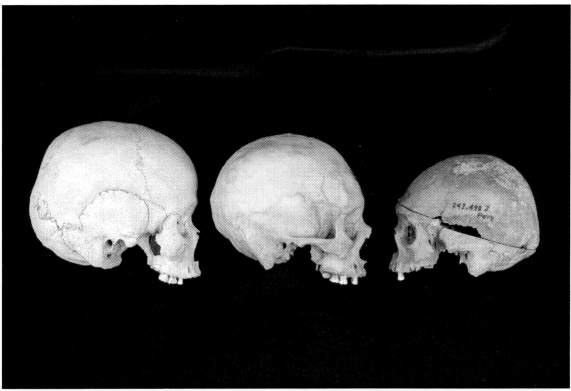

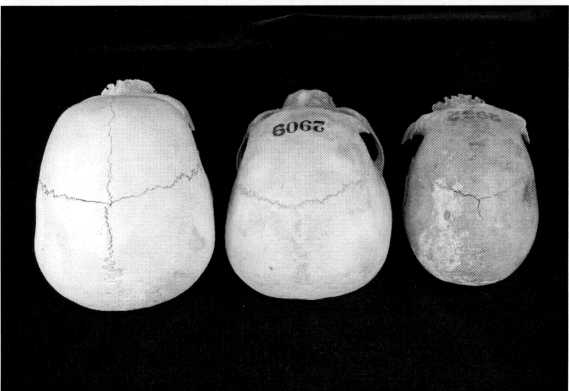

Figure 56a & b. Macrocephaly (1910cc), left; normal cranium, central; and microcephaly (910cc) from Peru. (NMNH 293381, left; 242549, center; 242498, right)

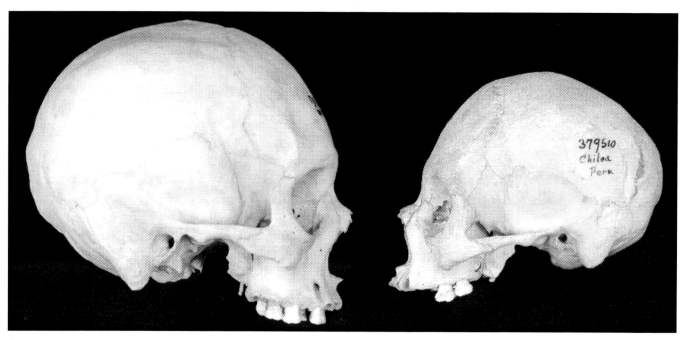

Figure 57. Normal cranium compared to an extreme microcephaly (485cc) from Peru. (NMNH 242549, left and 379510, right)

result of hyperpituitary growth. Zimmerman and Kelley 1982:23–26.

Microcephaly is a rare, neurological disorder where the circumference of the head/skull is smaller than the average for the sex, and age in the population (**Figures 56 & 57**). The general range of microcephaly is considered to be from 4–700ccs (compared to 11–1400ccs for normal crania) (Zimmerman and Kelley 1982:22). Microcephaly may be congenital or it may develop within the first few years of life. Infants may be born with normally sized heads that later fail to develop. Microcephaly is due to failure of the brain to grow. Cranial sutures should not have premature closure to be considered true microcephaly versus malformation due to stenosis. The face develops normally, resulting in a proportionally enlarged face relative to the rest of the cranium, producing a receding forehead with a small vault. Compare this morphology to artificial cranial deformation due to cradle-boarding, wrapping and other devices for cultural purposes (see **Figures 58–60**). Life expectancy for microcephalics is low. Barnes 1994:157–159; Krauss et al. 2003; McKusick 1969; Nyland and Krogness 1978; Zimmerman and Kelly 1982:21–23 (Basutu microcephalic illustrated in Figs. 3 & 4).

A significant number of references to cranial deformation in cultures throughout the world are in the literature, discussing the morphological design as well as cultural and physiological significance (**Figures 58–60**). The crania illustrated here are only to show the amazing plasticity of the human cranium when intentionally restricted or bound not

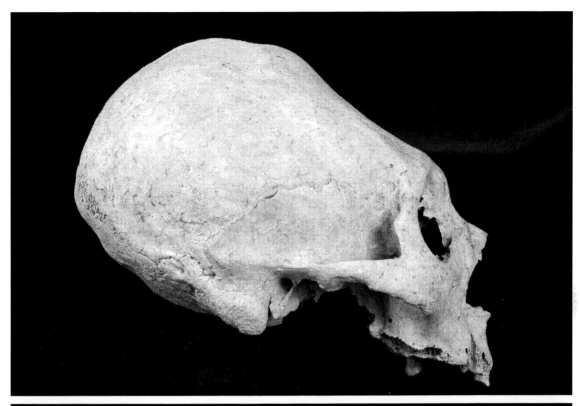

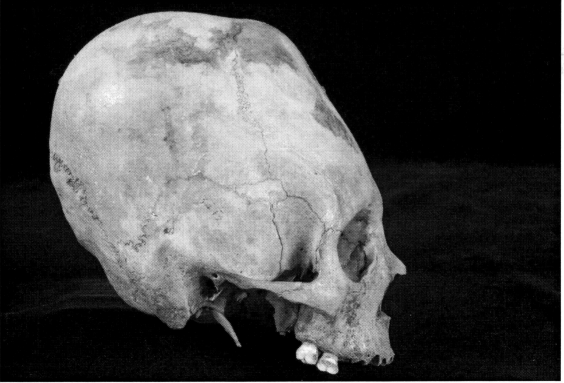

Figures 58, 59 & 60. Types of artificial cranial deformation. (Fig. 58–NMNH 378586; Fig. 59 - NMNH 276106; Fig. 60–NMNH 293689)

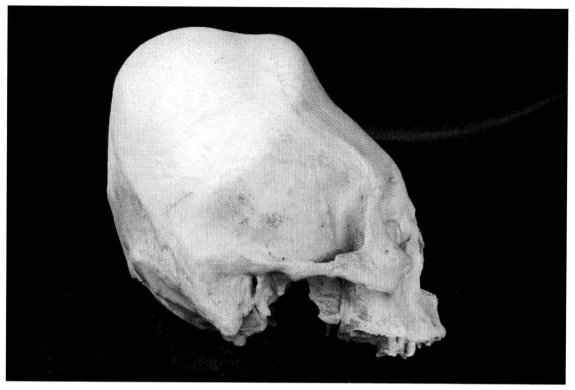

Figure 60.

to encompass the complete range of this practice and its results. Illustration of eleven different types of head wrappings found in mummies from South America is presented in Allison and Gerszten (1982). Allison et al. 1981; Aufderheide and Rodriguez-Martin 1998:34–36; Dean 1995; Gerszten 1993; Hrdlicka 1912; Moss 1958; Stewart 1941, 1973; Ubelaker 1999:96–99.

Sharp instrument defects on the bone are generally identifiable by a well defined margin where the blade or sharp edge has cut into the bone (Figure 61a). Pronounced chopping (**Figure 61a**) may be seen in butchering and defleshing procedures, deeper as the blade removes large portions or wedges of the bone, as represented in this illustration. Fiorato et al. 2001; Ortner 2003:137–142.

Gunshot wounds are sometimes misidentified as lesions and vice verse (**Figures 62, 63**). Gunshot wounds, in the present case, an exit wound in the left side of the frontal bone with radiating fractures, will almost always exhibit at least one beveled margin if it passes through bone. The beveled margin, which can be seen here, reflects the direction that the bullet was traveling through the body and is characterized by an irregular/porous/jagged defect encircling the gunshot wound. In the present case the bullet entered the person's head from the right

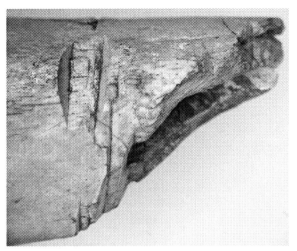 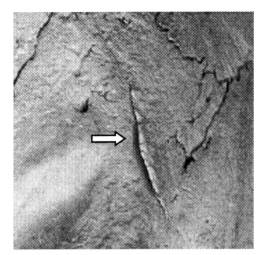

Figure 61a & b. Perimortem chopping injuries to a long bone (left) and slicing injury to skull above right mastoid process. (Right photo courtesy Greg Berg, Killing Fields, Cambodia)

mandible and traveled upward and to the victim's left, exiting the frontal bone. Radiographs may reveal evidence of small metallic bullet or casing fragments in or around the gunshot wound, or scattered metallic fragments in other parts of the skull as a result of the bullet breaking up as it traveled through the head. Remote fractures may also be revealed in the radiographs where normal visual observation is not possible (e.g., the orbits, sphenoid, internal regions of the cranial base and endocranial regions). The number, frequency (sequence of injuries) and directionality of gunshot wounds in bone, especially those in the skull, can be interpreted based on the presence of internal and external beveling, as well as the pattern and termination of radiating and concentric fractures. For a description and biomechanics of bone associated with gunshot wounds in bone, refer to Berryman and Gunther (2000), Berryman and Symes (1998), Di Maio (1999); Dixon (1982), Smith et al. (1987, 1993). Grandmaison et al. 2000; Mann and Owsley 1992; Quartrehomme and Iscan 1999; Ross 1996; Spitz 1980.

There is a broad spectrum of literature concerning trepanation to the cranium in prehistoric, historic and modern populations, and therefore review of these discussion are not necessary for this volume. To present this change to normal skeletal morphology, the figures presented here illustrate the scraping method (**Fig. 64**) and the cutting method (**Fig. 65**) of trepanation. They also illustrate both perimortem activities as well as the healed forms of this surgery, and in particular, **Figure 65** shows the typical undulating pattern of healing on the ectocranial table from the laying back of the scalp and damage to the periosteum during surgery. For an extensive overview and illustrations

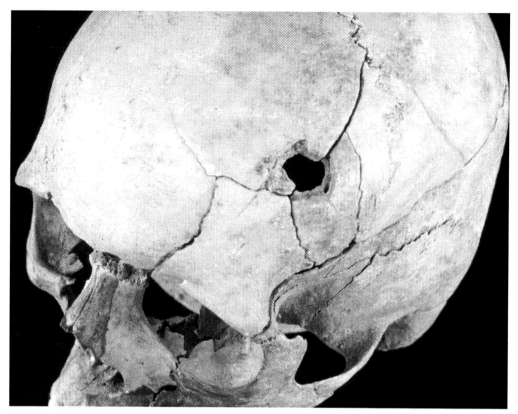

Figure 62. Gunshot wound to the frontal bone. (Historic North Korea)

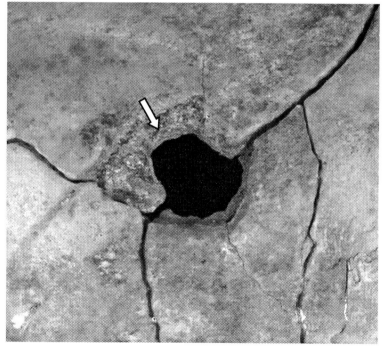

Figures 63. Close up of exit gunshot wound in left frontal bone; note beveling/cratering (arrow) encircling the defect.

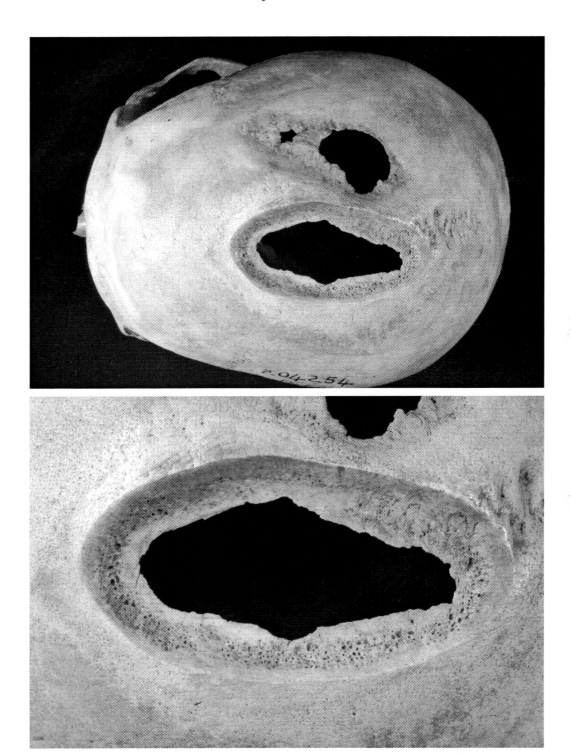

Figure 64a, b & c. Trephinations (trepanation) of the skull. Superior view of cranium and close-up of perimortem surgery by scraping (b) and close-up of a significantly healed trephination (c). (NMNH 204254)

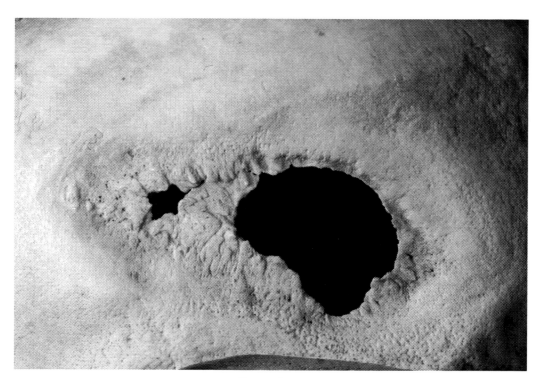

Figure 64c.

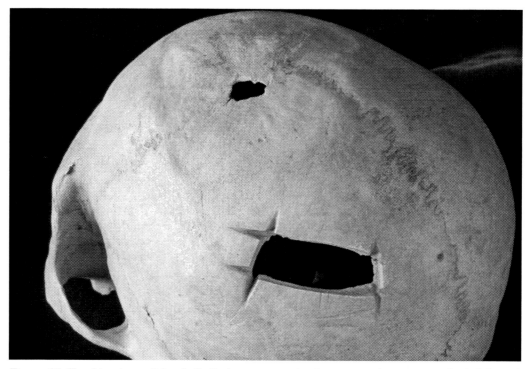

Figure 65. Trephinations of the skull. Perimortem cutting/sawing trephination on the left lateral parietal with cutmarks around the periphery of the surgery. The healed trephination located near bregma has an undulating area around the healed opening. This is remodeled bone from the lifting of the scalp and periosteum during the surgery. (NMNH 293778)

on the different types of trephination, see Lastres and Cabiese (1960) as well as Aufderhide and Rodriguez-Martin (1998:31–34) and Ortner (2003:169–174) and there is a series of 42 slides of trepined crania from the San Diego Museum of Man (Rogers and Merbs 1980). Margetts 1967; Rifkinson-Mann 1988; Stewart 1958; Tyson and Dyer 1980.

Trephination is an ancient technique of removing a portion(s) of the cranium by scraping (**Figure 64a & b**), drilling, or sawing. Although rare in North America, this practice is quite common in prehistoric skulls of South American (in particular, ancient Andean populations) as well as many other groups throughout the world (John Verano Pers. Comm. 1989). Webb (1988) reported on two skulls from Australia that may represent the first such cases from that continent. Care must be exercised when trying to differentiate lytic lesions (e.g., infectious diseases, tumors), axe and sword wounds, and blunt trauma from trephinations in the skull. Look closely for cut marks and abrasions around the defect that would indicate trephination. Healed trephinations will exhibit rounded margins, scarring, tiny spicules or roughened (possibly porous) areas surrounding the defect where infection or healing has occurred. In this example, one of the defects has not fully remodeled, as evidenced by visible diploe and relatively smooth (unscarred) margin. For additional background and references and illustrations of healed, healing and perimortem trephination, refer to Ortner 2003:169–174 and Aufderheide and Rodriguez-Martin 1998:31–34. Lisowski 1967; Sanan and Haines 1997; Shaaban 1984; Steinbock 1976:29–35; Stewart 1958; Velasco-Suarex et al. 1992.

Scalped skulls, in comparison, will exhibit patterned cuts that usually circumscribe the entire cranium. Typically the cuts extend across the middle of the frontal bone from temporal to temporal. Also look for deep or superficial cuts and scratches on any raised area of the skull where the scalp was removed by cutting through the muscles along the temporal lines, zygomatics, supramastoid crests, and mastoids. Steinbock 1976:24–30.

Rodent gnawing (**Figure 66**) may also be present in the skull as well as any of the long bones. Although it is sometimes difficult to distinguish rodent gnawing from defleshing marks (e.g., cultural practices before burial), the former are usually located above the eye orbits, nasal area, zygomatic arches, and mandible. Rodent gnawing appears as short but deep, *parallel* cuts localized to raised areas of the bone. Rodents can destroy entire bones. Wells 1967:10–11; Steinbock 1976; Ubelaker 1999:105.

Erosion and pitting (a) of the mandibular condyle—pitting and erosion of the condyle are common findings in all populations although the severity varies greatly from one group to another generally due to

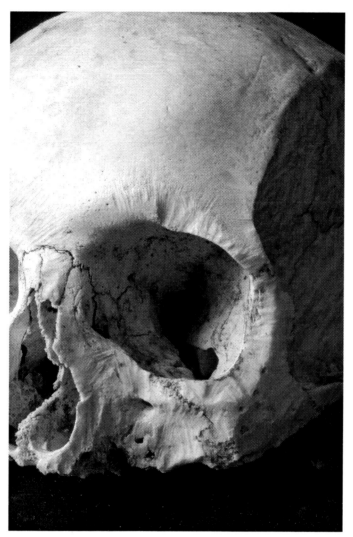

Figure 66. Rodent gnawing on bone. Notice the parallel striations on the bone surface from the rodent's incisors. (NMNH)

differences in the speed and severity of attrition. Porosity appears as a localized area(s) of bone loss, usually with a sharply defined margin. The common sites for such destruction are noted in **Figure 46**. The dotted line denotes bone loss due to erosion or, simply, the wearing away of the joint surface. Alteration of the mandibular condyle may range from a very small area of pitting (porosis) to complete destruction of the articular surface. Diet, mechanical factors related to chewing, dental wear, caries (cavities), disc abnormalities, facial morphology, and abscesses all contribute to destruction of joint. Richards 1987, 1988; Richards and Brown 1981; Ryan 1989.

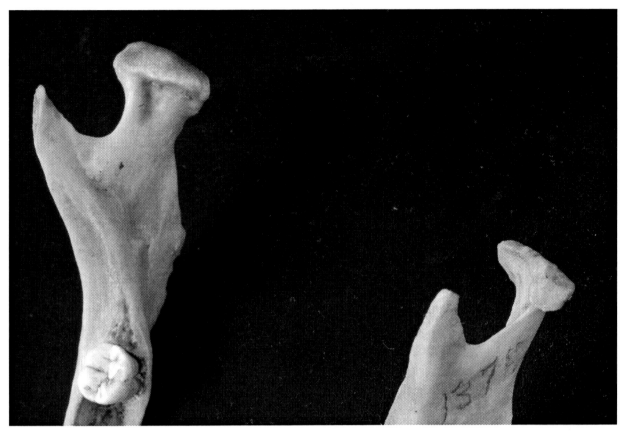

Figure 67. Pathological changes (OA) of the temporomandibular joint (TMJ) (Markowitz and Gerry 1950) of the mandibular condyles. (NMNH-T 1374)

Erosion (a) and osteophytes (b) of the mandibular condyle–when looking for erosion of the condyles compare the shape and size of both heads for symmetry. There may be slight asymmetry in the condyles that is developmental or genetic and not representative of a pathological condition. Slight erosion is a common finding in most populations while severe forms appear to be age-related (elderly) or the result of trauma or chewing stress.

VERTEBRA

The atlas may exhibit marginal osteophytes and surface porosity on the articular facets or dens (odontoid process (**Figure 68**). In some instances an elderly individual will exhibit eburnation (polishing) of the anterior surface of the dens and ossification of the apical ligament (most superior portion of the dens). All of these changes are consistent with OA/DJD. Erosion of the dens, however, is one of the criteria

Figure 68. Osteoarthritis of the second cervical vertebra (axis). (NMNH-H)

associated with rheumatoid arthritis (RA). (In advanced cases of RA the entire dens may erode and fracture resulting in death of the individual.) The first cervical vertebra may also exhibit corresponding changes. OA is a common finding although unequivocal cases of RA in archaeological specimens are rare. Bland 1994; Sager 1969.

The most common finding of OA in the cervical vertebrae is marginal osteophytes (osteophytosis) of the anterior and posterior inferior margins of the body (**Figure 69**). Although the anterior border may be slightly irregular in normal vertebrae, osteophytes appear as bony extensions and spicules projecting inferiorly. Marginal osteophytes and surface osteophytes on the articular facets may also be present (**Figure 70**). In more severe cases, the vertebral body may exhibit macroporosity and distortion of the endplates. Bland 1994; Lestini and Wiesel 1988; Ortner 2003:555–558; Rogers et al. 1997; Sager 1969.

Note the irregular margins encircling both the superior and inferior facets, as well as moderate-severe pitting/porosity (what some refer to as "macroporosity") of the superior facets. These features are characteristic of any joint in the body affected by degenerative joint dis-

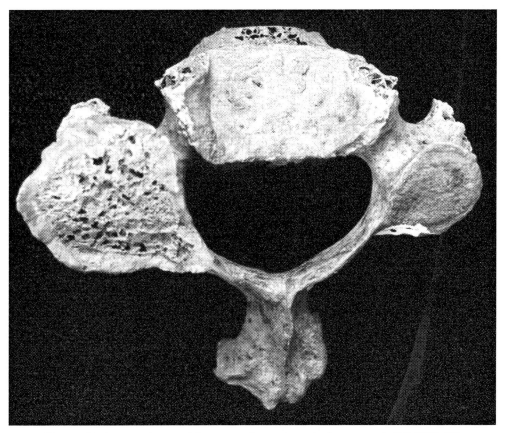

Figure 69. Osteoarthritis of the cervical vertebrae (inferior view). (NMNH–H)

ease/osteoarthritis.

"Hangman's fracture" traditionally refers to traumatic spondylolisthesis of the axis with separation of the body and pedicles in judicial hangings due to hyperextension of the upper cervical spine and distraction resulting in lethal damage to the spinal cord. Fracture of the axis, most commonly transversely across the entire vertebra posterior to the dens, can result from a variety of situations including hanging, motor vehicle accidents, a fall on the head, or being struck with an object (for example a wooden beam) (Junge et al. 2002; Smith et al. 1993; Williams 1975). James and Nasmyth-Jones (1992) examined the cervical vertebrae of 34 victims of judicial hangings and found fractures in only seven cases; six of the axis/C2 and one to C3. Hangman's fracture, therefore, may be less common than previously thought (**Figure 71**). Foreman 2001; McCort and Mindelzun 1990.

This is the form of bone growth that develops in degenerative joint disease. Note its resemblance to a bird's beak (**Figure 72**) and degree of curvature as it bridges the two vertebral bodies (see also **Figure 78**). Common in all populations. Hilel 1962.

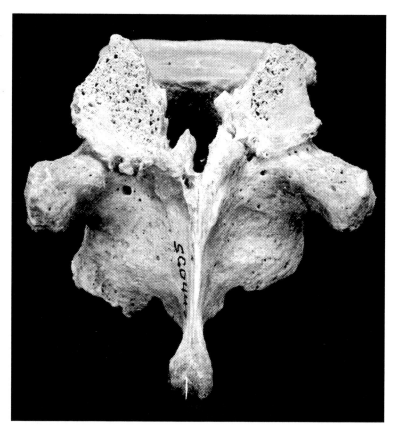

Figure 70. Osteoarthritis of the superior articular facets of a thoracic vertebra. (NMNH-H)

Although the etiology is unknown, DISH (**Figure 73**) is a common disease in middle-aged and elderly individuals with males affected 2:1 over females (Utsinger 1984). Although DISH frequently affects the spine, other peripheral skeletal sites may be involved and exhibit "whiskers" or "whiskering" periostosis (Burgener and Kormano 1991; Utsinger 1985), or irregular ossifications known as enthesophytes (spikes or projections) reflecting inflammation at the insertions of tendons and ligaments on the iliac crest, ischium, greater and lesser trochanters, trochanteric fossa, patella, calcaneus, ulna, and linea aspera. Aufderheide and Rodriguez-Martin 1998:97–99; Ortner 2003:559–560; Scutellari et al. 1995.

Criteria (Resnick and Niwayama 1976) for distinguishing DISH from ankylosing spondylitis include: (1) flowing calcification (resembling candle wax) and ossification along the anterolateral surfaces of at least four contiguous vertebrae, (2) relative preservation of intervertebral disc space in the involved segments, and (3) the absence of apophyseal joint (facets) bony ankylosis and sacroiliac erosion, sclero-

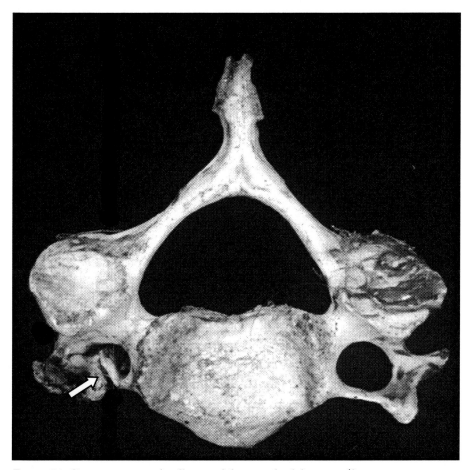

Figure 71. Compression and collapse of the vertebral formaen (foramen transversarium) in the axis (C-2) vertebra of a man who was hanged. Collapse of the foramen was the result of traumatic constriction of the noose; note that the left foramen is normal.

sis or osseous fusion (compare to **Figure 87** of an ankylosed spine). Utsinger (1985) revised the diagnostic criteria of Resnick and Niwayama as follows: continuous ossification along the anterolateral surfaces of at least two contiguous vertebral bodies, primarily in the thoracolumbar spine (broad, bumpy, buttress-like band of bone) and symmetrical enthesopathy of the posterior heel, superior patella or olecranon. Garber and Silver 1982.

Rarely ankylosing spondylitis and DISH coexist. An extensive radiographic survey of 8,993 persons over the age of 40 by Julkunen et al. (1973) revealed a standardized prevalence of 3.8 percent in men and 2.6 percent in women. Interestingly, the Pima Indians of Arizona showed an incidence of 34 percent in males and 6.6 percent in females (Henrard and Bennett 1973). Weinfeld et al. (1997), in examining 2,364 patients 50 years or older, found the highest incidence in Whites, fol-

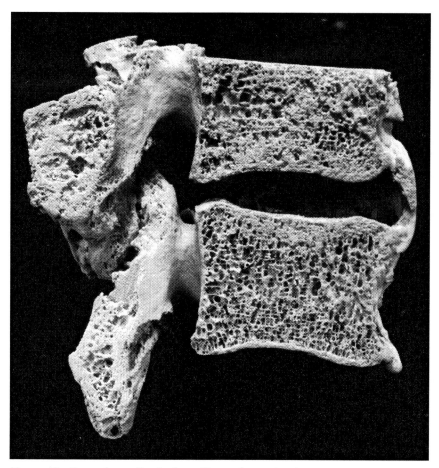

Figure 72. Osteophyte–"beak-shaped" new bone that buttresses weakened vertebral bodies in the elderly (osteoporosis) and/or trauma to the spine. (NMNH–H)

lowed by Blacks, Native Americans and Asians, indicating some population genetic factors. Ankylosing spondylitis is inherited as a Mendelian dominant trait with 70 percent penetrance in males and 10 percent in females (Sharon et al. 1985). There appears to be a metabolic component as well evidenced from Julkenen et al. (1971) findings that 25 percent of the survey patients in Finland with DISH also suffered from adult onset diabetes. Uncommon finding.

Note that an arrow flowing osteophytosis–thickened, flowing ossification is generally found along the right anterolateral aspect of the thoracic vertebrae. The left side of the spine is usually spared due to the presence of the aorta. Note that the "ribbon" (Scutellari et al. 1995) of ossification is clearly visible, extends across at least four contiguous vertebrae, and appears as if a thick, undulating, and smooth surfaced band of bone has been applied to the right side of the spine. Some of the thickened ribbon of bone is the ossified anterior longitudinal liga-

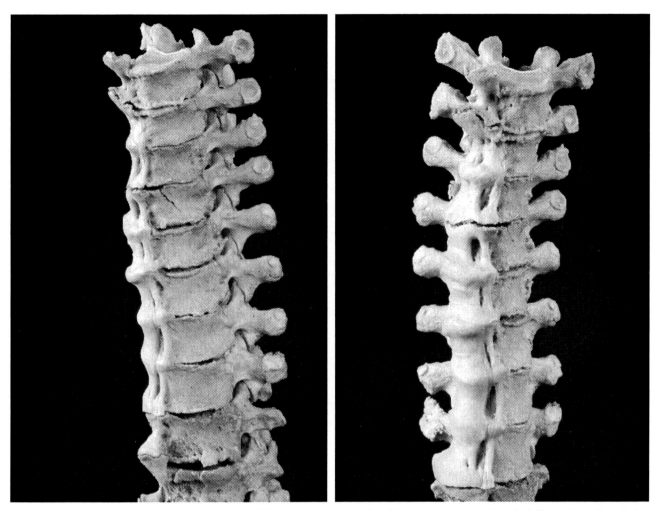

Figure 73a & b. Calcification and ossification of the anterior longitudinal ligament associated with diffuse idiopathic skeletal hyperostosis (DISH). (NMNH–T 236R)

ment. Aufterheide and Rodriguez-Martin 1998:97–99; Julkenen et al. 1971; Ortner 2003:559–560, 571–577; Rogers and Waldron 1995; Steinbock 1976:296–297; Utsinger 1985; West 1949.

McKern and Stewart (1957) reported that laminal spurs (**Figure 74**) were most often found in the thoracic vertebrae, then the lumbar and, with rare exception, the cervicals. These normal variants of the spine are often seen associated with increasing age or believed to accompany strenuous activity. Regardless of etiology, the authors have encountered large spurs in children. This condition can be scored nonmetrically either as present or absent, or as mild, moderate or severe. Some researchers have measuresd the lengths, but this practice is difficult to reproduce accurately. Common finding in all populations. Allbrook 1954; Davis 1955; Naffsiger et al. 1938; Shore 1931.

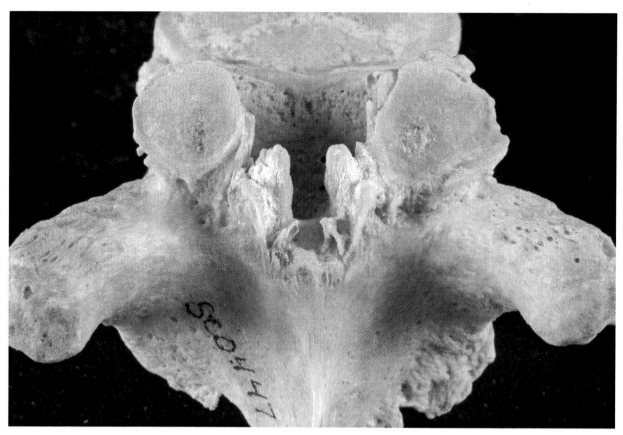

Figure 74. Laminal spurs–spikes along the superior border of the "V"-shaped neural arch (lamia) that serve for attachment of the inferior ligamentum flavum (Hilel 1959). (NMNH–H)

Compression fractures and collapse of the vertebral centra produce forward bending of the spine (Kyphosis) manifested as a hump back, most often observed in the elderly (**Figure 77**). The varying loss of centrum height (see **Figures 75 & 76** for normal vertebral development in children) most frequently results from collapse of the superior endplate (which possibly can be also associated with Schmorl's nodes) (see **Figs. 83 & 84**), not the inferior plate. As the intervertebral disc protrudes into the endplates, the body of the vertebra gives way and collapses over time. Arthritic response to the exuding of the intervertebral disc anteriorly and laterally will also produce osseous spicule formation or osteophytes along the rim of the centrum (see **Figure 84**). Common finding in the elderly. Albright et al. 1941; Cummings et al. 1985; Dolan et al. 1994; Frymoyer et al. 1997; Horal et al. 1972; Raisz 1982; Tehranzadeh et al. 2000; Twomey & Jaylor 1988; Ubelaker 1999:84–87.

Rounded or radiating finger-like projecting bony growth(s) commonly found on the rim and vertebral bodies of older individuals.

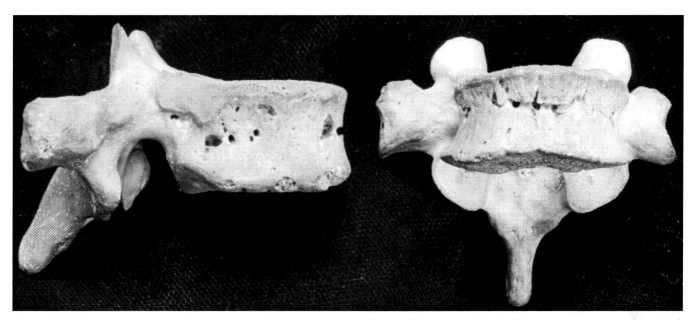

Figure 75. Normal nutrient pit for passage of the nutrient vessels into the centrum/body. (More prominent in subadults) (NMNH–H)

Figure 76. Normal billowing ("sunburst") of vertebra, especially prominent in thoracic and lumbar vertebrae, for attachment of the endplates and epiphyseal rings. (NMNH–H)

Figure 77. Compressed vertebra (collapsed vertebra, wedged vertebra)–results from bone loss accompanying old age (osteoporosis), infection, or trauma such as a compression injury. (NMNH–H)

However, these formations are not age specific, being found in association with younger individuals who have either sustained trauma to the spine or due to any number of spinal diseases such as tuberculosis, mycotic infections, etc. Osteophytes will result in small to large rounded protrusions at two or more adjacent vertebral bodies, appearing as small, raised irregularities along the margins or midsections of the centra and that gradually become larger, having a "parrot-beak" appearance until bridged fusion of the bodies occurs. When two vertebrae become fused (ankylosed), the large osteophytes may be referred to as kissing or bridging osteophytes (**Figures 72, 78**). Common finding. Hough and Sokoloff 1989; Stewart 1958.

Pronounced *spon-di-loli-sis* (Thomas 1985) separation of the vertebral body from the posterior vertebral arch, usually at a junction known as the pars interarticularis or isthmus (**Figures 79, 80**). Some researchers believe this condition to be congenital in origin while others state that stress plays a major role in causing the neural arch to separate, or basically a stress fracture and nonunion (see Merbs, 1983;

Figure 78. Osteophytes (osteophytosis, spondylitis deformans, spondylosis deformans). (Dieppe et al. 1986) (NMNH–H)

Newman and Stone 1963; Stewart 1953, 1956). Spondylolysis is seen in between 3 percent and 10 percent of the adult population (Collier et al. 1985), but occurs primarily in children with an increasing incidence until 20 years (Ohmori et al. 1995; Roche 1949; Roche and Rowe 1951, 1953) and an arrest of occurrence after age forty years (Stewart 1953; Wiltse 1962).

Activity, especially shearing stress, involving the lower spine does seem to contribute to the presence of this trait. Wynne-Davies and Scott (1979) reported that the frequency of spondylolysis is highest in athletes, and the highest risk groups are Alaskan Native American and a first-degree relative with spondylolysis at a risk level of 50 percent in development of spondylolysis (see Gerbino and d'Hemecourt, 2002 for discussion of lumbar involvement due to football sports). The frequency of spondylolysis is found as high as 40 percent in Eskimos

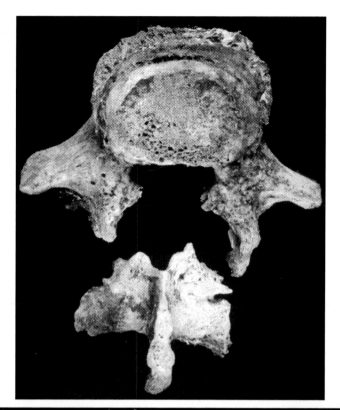

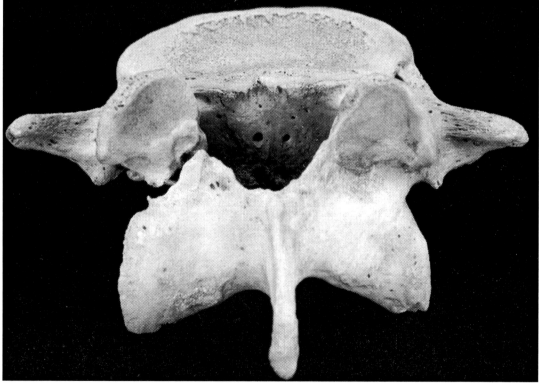

Figures 79 & 80. Bilateral (79) and unilateral (80) spondylolysis (separate neural arch, Stewart 1931, 1956; bipartite lumbar, Grant 1972). (Fig. 79–NMHH Egypt; Fig. 80–NMNH-H)

(Stewart 1953; Boyer et al. 1994). As it relates to activity, the incidence of spondylolysis in divers was found to be 63 percent and in gymnasts, 32 percent (Gerbino and Micheli, 1995). Gerbino and Micheli (1995) contend that spondylolysis is the result of mechanical forces–primarily repetitive hyperextension of the lumbar spine–on the pars interarticularis (Omey et al. 2000). Cyron and Hutton (1978:238) hypothesized that spondylolysis is more likely to occur in young people because they frequently engage in strenuous activity at a time when their intervertebral discs are more elastic and their neural arches may not be completely ossified (see Fehlandt and Micheli 1993 for report on lumbar stress fractures in ballet dancers, and Brukner et al. 1996 study of 180 sports athletes). Thieme (1950) hypothesized that spondylolysis results from trauma during the developmental stage to final fusion. The lack of normal repair callus at the lesion, even in unilateral instances where movement of the "separating" arch is held at a minimum, suggests against postfusion breaks. Garth and Van Patten (1989) reported a case of bilateral fracture of the lumbar laminae, not all, but near, the pars interarticularis that showed no bony union after two months. Friedman and Micheli (1984) reported spondylolisthesis due to stress concentration on the neural arch from scoliosis surgery.

Most commonly, the fifth lumbar is affected, but the fourth and third may also show this trait. Spina bifida is a common finding associated with separate neural arch defects (Rowe and Roche 1953). Eisenstein (1978) found the defect in 17 skeletons (3.5%) of 485 examined; 1 example in L3, 4 in L4, and 13 in L5. Libson et al. (1982) reported that of 1598 patients with lumbar spondylolysis, only two individuals had lesions in the upper three lumbar vertebrae. However, spondylolysis in the lower two lumbar vertebrae is more common. Rowe et al. (1987) found the defect in the lower lumbar in 2 percent to 10 percent of active young individuals in the United States. The male to female ratio is 2:1 from age six to adulthood (Fredrickson et al. 1984), and the condition has never been reported as present in newborns. The youngest individual on record with spondylolysis is a 3.5-month-old infant (Seitsalo et al. 1988).

Although the etiology of spondylolysis remains unresolved, including genetic and congenital factors, (Blackburne and Velikas 1977; McKee et al. 1971), repeated stresses and trauma to the lower back appear to play a major role in its development (Kono et al. 1975). Spondylolysis is a common finding and is strongly associated with spina bifida and repeated hyperextension of the lower spine (Barnes 1994:46–53). Separation of the neural arch may be incomplete, complete, unilateral, or bilateral. Care should be exercised not to confuse spondylolysis with postmortem breakage, which is common at this site. Aufderheide and Rodriguez-Martin 1998:63; Barnes 1994:46–50, 259–265; Bradford 1978; Bridges 1989; Eisenstein 1978; Floman et al. 1987; Fredrickson et al. 1984; Griffiths 1981; Harris and Wiley 1963; Hitchcock 1940; Hoppenfeld 1977; Lamy et al. 1975; Laurent and Einola 1961; Letts et al. 1986; Libson et al. 1982; Lowe et al. 1987; Miles 1975; Nathan 1959; Newman and Stone 1963; Ortner 2003:147–149; Pecina and Bojanic 2004; Ravichandran 1981; Rosenberg et al. 1981; Stewart 1931,1953 and 1956; Thieme 1950; Troup 1976; Wiltse 1972; Wiltse et al. 1975.

If the vertebral body slides forward on the sacrum (which may be detected by the presence of bridging osteophytes between L–5 and S–1), the condition is known as spondylolisthesis and presents a more complicated condition (Congdon 1931; Merbs and Euler 1985; Pedersen and Hagen 1988; Ruge and Wiltse 1977; Thieme 1950; Wiltse 1962; Wiltse et al. 1975). Look for (1) fissures below the superior articular facets with or without evidence of healing (callus) that may or may not impinge on the neural canal; and (2) unusually large/hypertrophied or asymmetrically shaped inferior articular facets on the side of the spondylotic defect. Uncommon to common finding depending on the population and sample studied. Bradford 1978; Farfan et al. 1976; Lester and Shapiro 1968; Merbs 1995, 1996, 2002; Miki et al. 1991; Nathan 1959; Neugebauer 1976; O'Beirne and Horgan 1988; Porter and Park 1982; Ruge and Wiltse 1977; Saifuddin et al. 1998; Waldron 1992, 1993; Willis 1924, 1931; Wiltse et al. 1976.

This is a congenital malformation of unknown etiology (**Figures 81, 82, 85**) resulting in incomplete closure of vertebral neural arches (see discussion in Barnes 1994:117–125 on the embryonic developmental problems which produces these anomalous conditions). When examining dry bone (no soft tissue or a patient history), it is extremely difficult, if not impossible, to determine if the condition was occulta (asymptomatic), cystica, or aperta. Cystica implies that the skin was involved and the lesion became cystic, while the term *aperta* refers to an open lesion (Strassberg 1982). (Dickel and Doran 1989 present an interesting case of a 15-year-old child dating to 7,500 years BP diagnosed as having aperta). The most common site of neural arch deformity is S1 with a reported incidence of 9 percent in females and 13 percent in males (Cowell and Cowell 1976), but the entire sacrum can be involved (see **Figs. 107 & 108**). Care must be exercised when examining and interpreting archaeological or ancient dry-bone specimens as herniation of the spine can occur in even mild cases of spina bifida (James Vailas pers. comm. 1989). Barnes 1994; Ferembach 1963; Saluja 1988.

Schmorl's depressions can be a circular, linear or combination of

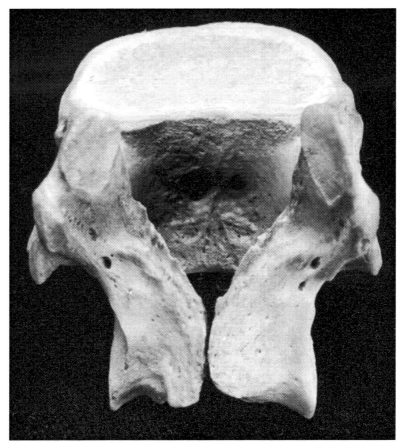

Figure 81. Bifid neural arch (spina bifida, split or cleft arch). (NMNH–H)

the two, depressed lesion(s), usually with a sclerotic floor in either of the centra endplates (**Figures 83, 84**). In some cases only a small circular depression will be present in the center of the centrum—such lesions should also be scored as present. Schmorl's depressions result from herniation of the nucleus pulposus, the partially liquid central portion of the intervertebral disc. This herniation of the nucleus can take place in any direction but will bulge where the bone (frequently the endplate) or annulus fibrosus is weakest.

Schmorl's depressions are common findings in the elderly and result from degenerative disc disease. However, the presence of such nodules/depressions can be seen in adolescents but is uncommon (only 2% of all herniated discs occur in children and adolescents (Bunnell 1982)). Schmorl's depressions in subadults result from trauma from such activities as a fall from height, heavy lifting, trauma during physical exercises, and similar activities (Bolm-Audroff 1992). Approximately 2 percent of all individuals, however, will have a clinically significant herniated nucleus pulposus at some time in their lives

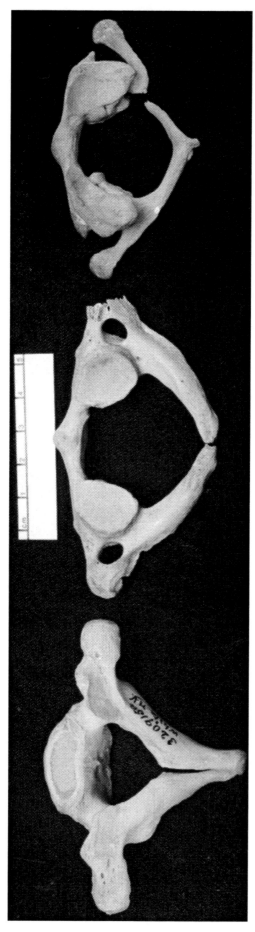

Figure 82. Cleft/bifid arches in thoracic, atlas (C-1) vertebrae. Note that the cleft is along the midline in one C-1 and unilateral in the other. Barnes 1994:120–124. (NMNH–H)

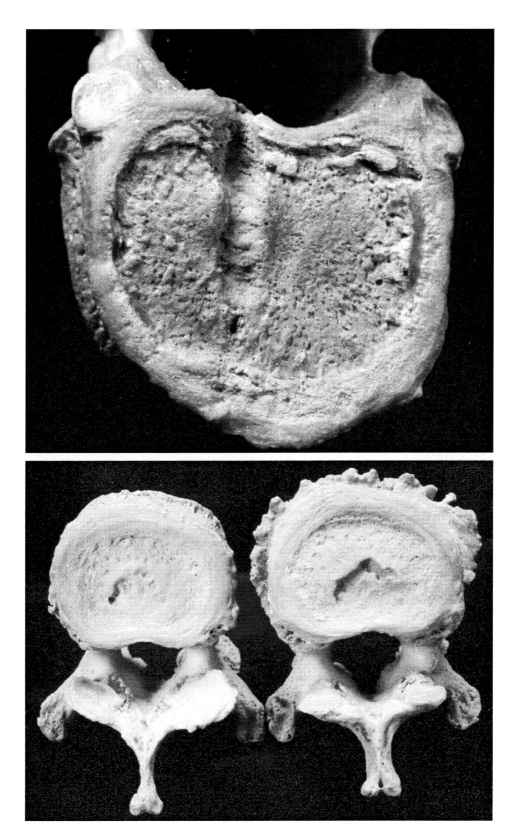

Figures 83 & 84. Schmorl's depression (Schmorl's nodule, node, or cavity, disc hernia-tion, cartilaginous node). (NMNH–H)

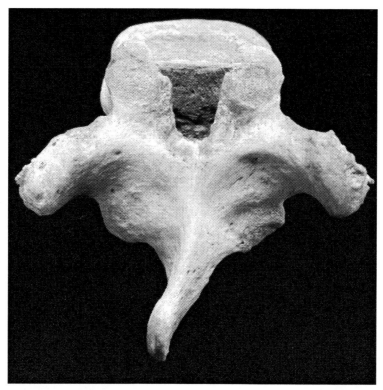

Figure 85. Deviated, but otherwise normal, spinous process (lumbar vertebra shown). (NMNH–H)

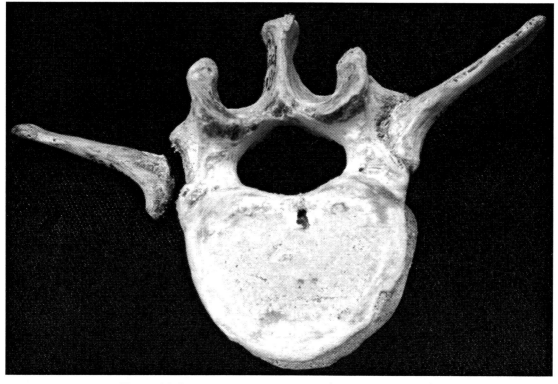

Figure 86. Separate transverse process. (NMNH–H 317688)

(Frymoyer 1988).

Schmorl's depressions are common (2% to 76%) in all populations (Yochum and Rowe 1996). When scoring its presence, the count can be based on either the number of vertebrae with the trait, how many total lesions are present in both the superior and inferior endplates, or the exact position of the lesions (e.g., between L–3 and L–4). For a good comparison of the frequency of herniated discs among populations see Thieme (1950). Bulos 1973; Campillo 1989; Gibson et al. 1987; Giroux and Leclerq 1982; Jayson and Dixon 1992; Lindblom 1951; Resnick and Niwayama 1978; Saluja et al. 1986; Schmorl and Junghanns 1932; Tehranzadeh et al. 2000; Weinstein et al. 1977.

A condition resulting in a divided transverse process (**Figure 86**). A smooth-surfaced articular facet is associated just inferior to the superior articular facet at the base of the mammillary tubercle. This is different from a lumbar rib which attaches to the centrum and has a separate transverse process to the neural arch. This developmental variant is the result of caudal shift in the developmental pattern of the spine and affects the formation of the lumbar vertebra. The authors have encountered only one unilateral case in a first lumbar vertebra. Rarely a transverse process, usually a lumbar vertebra, may fracture and separate (Hoppenfeld 1977). Rare finding. Barnes 1994:104–108.

Ankylosing spondylitis (AS) (**Figure 87**) is a chronic inflammatory disorder of unknown etiology (Borenstein et al. 1995) but has been linked (95%) to the histocompatability complex antigen HLA–B27 (Aufderheide and Rodriguez-Martin, 1998:102; see Gerber et al. 1977 for alternative view) and is identified as a Mendelian dominant trait (Sharon et al. 1985). The disease is predilected to males (70%, 10% in females as reported by Sharon 1985 to 90 percent in males as reported by Aufderheide and Rodriguez-Martin 1998). The prevalence of AS varies by population from absent in Black Africans and Australian aborigines to 4.2 percent in adult male Haida Indians (Masi and Medsger 1981) but Caucasian males are the most predilected for the disease (Aufderheide and Rodriguez-Martin, 1998:102). Contemporary populations show an incidence of about 1 per 1000 individuals (Resnick and Niwayama 1981) to 1 per 2000 (West 1949) with a male to female ratio varying from 4:1 to 10:1 (Resnick and Niwayama 1981; West 1949). The typical age of onset is 15 to 35 years but children may also be affected with juvenile-onset AS) (Jaffe 1975). Bony changes typically begin in the sacroiliac region (sacroilitis and subsequent ankylosis) and extend up the spine. As the spine becomes more involved it takes on an undulating contour ("bamboo spine") due to the development of extensive syndesmophytes between the vertebral bodies (Resnick 2002; Resnick and Niwayama 1981). In

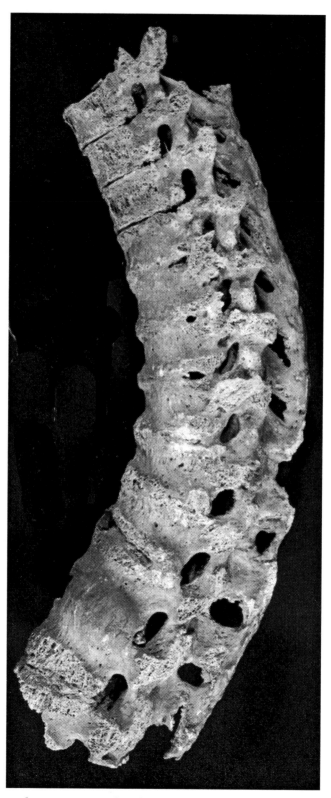

Figure 87. Bony changes associated with ankylosing spondylitis (AS), and degenerative joint disease (spondylosis) of the spine. Hough and Sokoloff 1989; Kahn 1984; Meisel and Bullough 1984; Ortner 2003:571–577; Wei Yu et al. 1998.

extreme cases, the ribs will also be involved in the osseous fusion (for a classic case of AS with rib involvement see Ortner 2003:576, Fig. 22.10). AS often results in ossification at tendinous and ligamentous attachment sites (enthesophytes) to bone and in some cases may be difficult to distinguish from DISH (see **Fig. 73a & b**). Calin (1985), Dieppe (1986), Long and Rafert (1995), Ortner (2003), and Resnick and Niwayama (1988). Also see Epstein 1976; Gibson et al. 1987; Giroux and Leclerq 1982; Schmorl and Junghanns 1932, 1971a & b; Steinbock 1976:294–298, 304–309; West 1949.

Ossified tri-radiate ligaments. Resulting in fusion of the ribs to the vertebrae are typical of ankylosing spondylitis. Uncommon to rare finding.

Syndesmophyte (endesmophytes). Vertically-oriented bone growth typical of AS that forms from within the *margins* (annulus fibrosus) of the vertebral bodies (Francois 1965). Common finding, depending on the population under study. Resnick and Niwayama 1988.

Osteophyte. Horizontally-oriented, rounded bone growth associated with degenerative spinal disease (osteophytosis), increasing age, and trauma. Common finding in all populations.

Spondylosis deformans (SD) is the most common degenerative disease of the spine that affects males much more frequently than females, is occupation related, and is found in nearly all individuals over the age of sixty (**Figure 88**). SD is characterized by small to large bridging osteophytes (arrows) that bulge at the level of the intervertebral disc and serve to reinforce the centra. The proper use of the term osteoarthritis refers only to degeneration and involvement of the apophyseal joints (articular facets) whereas osteophytosis refers to osteophyte formation along the vertebral bodies and degenerative disc disease (Hough and Sokoloff 1989). The following description given by Norman (1984) best illustrates the process of osteophyte formation commonly associated with old age:

> The early changes of spondylosis deformans affect the anterolateral margin of the vertebrae where the annulus fibrosus inserts. Tearing of the fibers weakens the annulus. The restraints on the nucleus pulposus are lost, and the disk protrudes forward. Further stress will lift the anterior longitudinal ligament from the vertebral body, and a buttress of periosteal new bone fills in the area of separation. The osteophytes form in a horizontal direction and curve to bridge the intervertebral disk space. . . .

Many researchers attribute osteophytosis to many years of wear and tear that necessarily accompanies old age (Keim 1973). Trauma, heavy physical stresses to the spine, and obesity, however, may also result in osteophyte formation. Dieppe et al. 1988; Hough and Sokoloff 1989; Jaffe 1975; Moskowitz et al. 1984; Ortner and Putschar 1985;

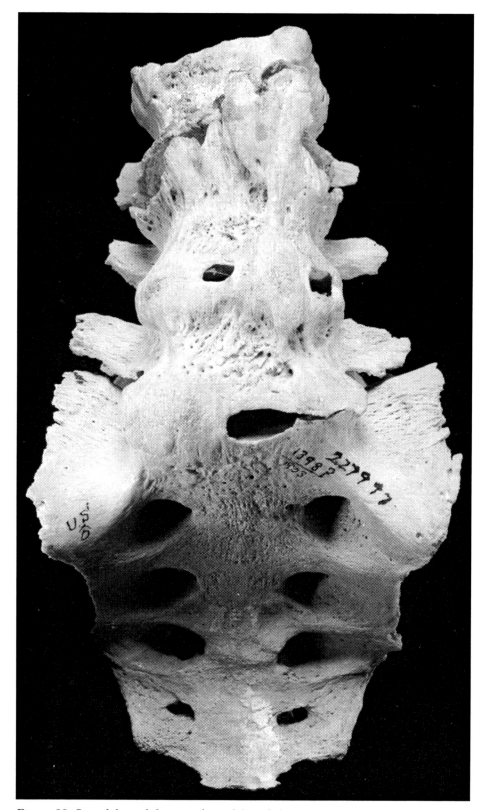

Figure 88. Spondylosis deformans (spondylitis deformans, degenerative hypertrophic spondylitis, osteoarthritis, degenerative spondylosis) of the spine. (NMNH–H 227977)

Steinbock 1976:287–289; Stewart 1958; Trueta 1968.

An aortic aneurysm may result in large, scooped-out lesions of the vertebral bodies, sometimes accompanying syphilis (**Figure 89**). These are caused by positive charges from the exterior surface of an expanded aorta, the eroded portions of the vertebral bodies resulting from the expansion of the aorta. Typically, males over the age of 50 are affected (Keim 1973). The lesions are restricted to the left antero-lateral surface of the spine. Distinguishing aneurysmal from tubercular erosion in an isolated vertebra is difficult and may require the assistance of a radiologist. Rare finding in most populations (infrequent finding in cadaver and necropsy collections). A good review of aneurysms from various diseases and their bony effects as well as illus-

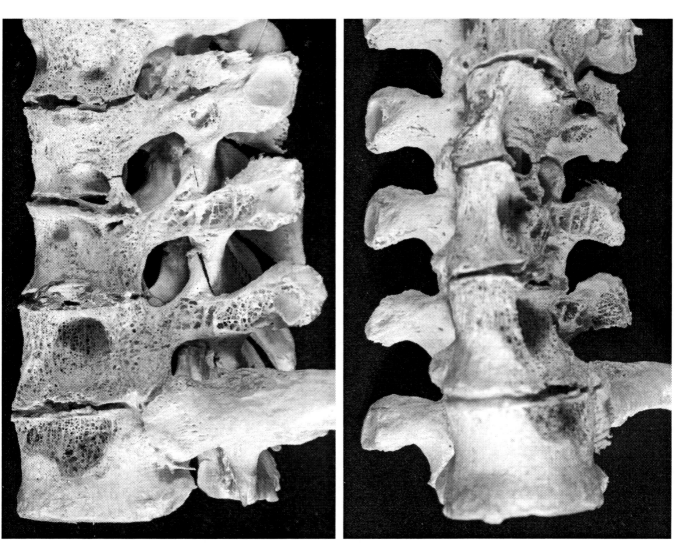

Figure 89a & b. Aortic aneurysm of the vertebrae.

tration of involvement of vertebrae and sternae is found in Aufterheide and Rodriguez-Martin 1998:78–81. Choi and Harris 2001; Diekerhof et al. 2002; El Maghraoui et al. 2001; Leung et al. 1977; Mii et al. 1999; Ortner (2003:356 Fig. 13–11 & Fig. 13–12); Semrad et al. 2000; Zimmerman and Kelley 1982:73–74.

Note that the drawing depicting a normally curved spine (figure left to right), kyphosis (humpback), and lordosis (swayback) are viewed from the side, while that of scoliosis is seen from the front (**Figure 90**). These conditions can only be detected by articulating the vertebrae in the approximate anatomical position (by aligning the articular facets). Remember that each intervertebral disc accounts for approximately 0.5 to 1.0 cm space (in young to middle-aged individuals) between adjacent vertebrae. Kyphoscoliosis, forward and to the side bending, is a common finding in the elderly. The dowager's hump of the elderly, usually most marked in females, is a common form of kyphosis. With increasing age, the vertebrae lose bone mass (osteoporosis), collapse, remodel, and become anteriorly wedge-shaped (see **Fig. 77**). Physiological scoliosis in which the vertebral bodies are greatly distorted and asymmetrical is a rare finding (**Figures 91, 92**). A number of etiologies may result in scoliosis due to loss of bone and support by lytic bone destruction including tuberculosis, non-specific osteomyelitis, and mechanical fatigue from acute trauma, osteoporo-

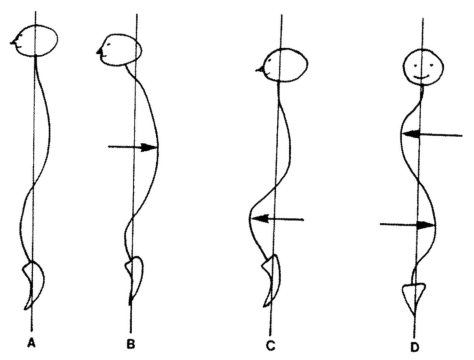

Figure 90. Three common forms of spinal deformity.

sis, or osteoid osteoma (Haibach et al. 1986; Savini et al. 1988; Swank and Barnes 1987). Barnes 1994:38–40, 59–71; Benzel 2001; Davis 1955; Zink et al. 2001.

Scoliosis is defined as "one or more lateral-rotatory curvatures of the spine" (Keim 1972). Idiopathic or genetic scoliosis accounts for approximately 70 percent of all scoliosis and is divided into infantile, juvenile, and adolescent. Scoliosis is a common finding in the elderly, especially women (postmenopausal or senile osteoporosis) above 65 years of age. This form, however, is due to osteoporosis and collapse of the vertebral bodies and is not congenital in origin. Moderate to

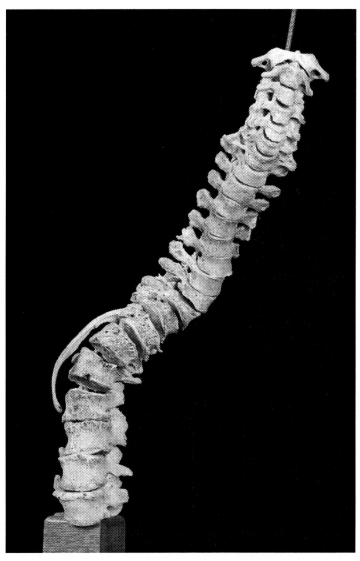

Figures 91 & 92a & b. Scoliosis of the spine (lumbar used as example). Fig. 91–NMNH-T 1636, Fig. 92–NMNHH)

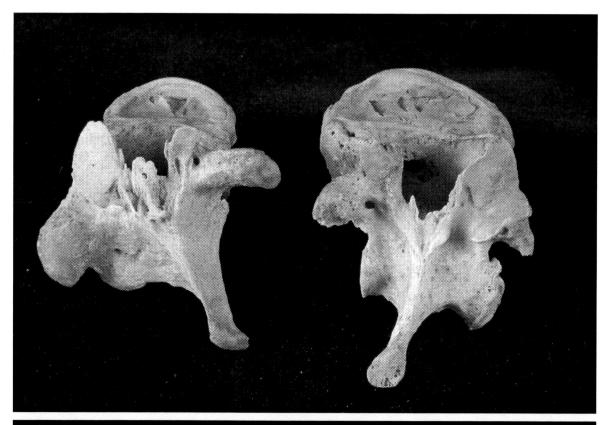

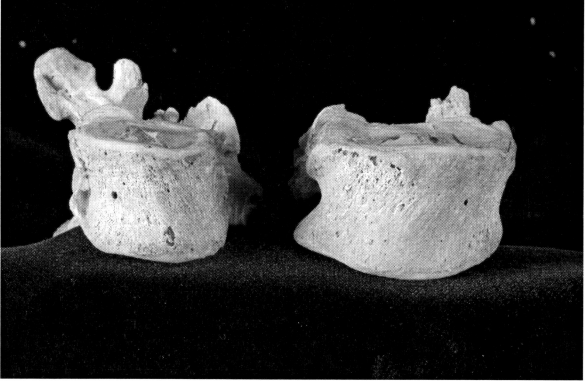

Figure 92a & b.

severe scoliosis in young individuals is a rare finding in most skeletal samples. Apley and Solomon 1988; Jackson 1988; Raisz 1982; Trueta 1968.

In examining a spine for physiological scoliosis look for twisted (asymmetrical) vertebral bodies (**Figure 92**), irregularly shaped and positioned articular facets, missing transverse processes and distorted and misshapen ribs. The upper half of the centra may appear to be shifted in one direction while the lower half is shifted in the opposite direction. When the vertebrae are correctly articulated the spine will spiral, sometimes to an extreme degree. Severe scoliosis is associated with a number of pathological conditions including cerebral palsy, Marfan's syndrome, neurofibromatosis, and Klippel-Feil syndrome (Klippel and Feil 1912; Prusick et al. 1985). Barnes 1994; Dickson 1985; Farkas 1941; Moe et al. 1978; Riseborough 1977; Simmons and Jackson 1979.

This vertebral form (**Figure 93**) is caused by the failure of fusion of the lateral halves of the vertebral body (Edelson et al. 1987; Epstein 1976; Moe et al. 1978; Pfeiffer et al. 1985). This rare condition is classified as a congenital defect of formation (developmental anomaly in the early embryonic period) (Muller et al. 1986) resulting in a variety of malformed centra (Barnes 1994:36–40) and is reported that it is more frequent in males than in females (Aufderheide and Rodriguez-Martin, 1998:62). The body of the vertebra may be cleft down the midline resulting in two hemivertebrae or thin stands of bone may transverse the cleft (Epstein 1976). Other bony changes may include widely spaced pedicles, malformed neural arches, and marked disarray of the posterior elements. The intervertebral disc may bulge into the defect. Butterfly vertebrae, so called because of their radiographic appearance when taken on the coronal plane, result in kyphosis (forward bending) due to the diminished height of the anteriorly wedged vertebral body. Note in the hemivertebrae in **Figure 94**, that two of the lumbar vertebrae failed to develop properly and are anteriorly wedged. Also note the large bridging osteophyte joining two of the vertebrae along their anterior surfaces. A police case involving a coronal cleft vertebra noted radiographically was initially suspected to reflect an abusive fracture in an infant (Aronica-Pollak et al. 2003). Aufderheide and Rodriguez-Martin (1998; p. 62, Fig. 4.9); Barnes 1994:60–67 (cites several paleopathological cases from New Mexico); Colquhoun 1968; De Graaf et al. 1982; Mann and Verano 1990; Taylor and Resnick 2000; Yochum and Rowe 1996; Zimmerman and Kelley 1982:30.

Tuberculosis in the spine (Pott's disease, vertebral tuberculous spondylitis) as is represented in this illustration shows the typical cav-

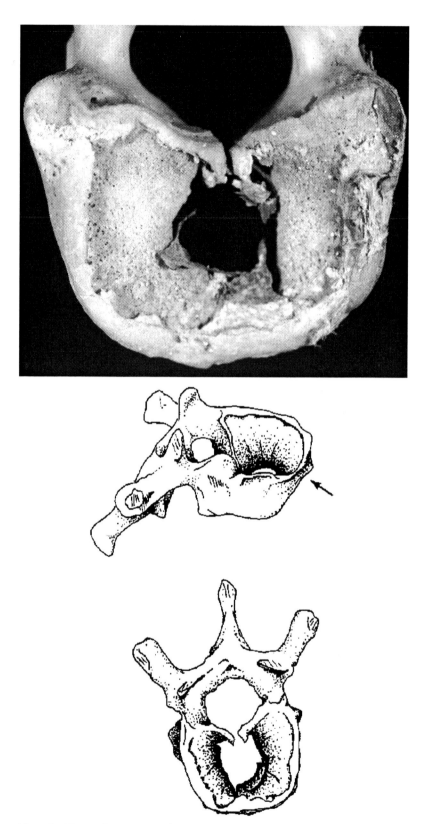

Figure 93a & b. "Butterfly" vertebra (sagittal cleft vertebra, drawings of oblique and superior view). (Vertebra from Peru, courtesy of John Verano)

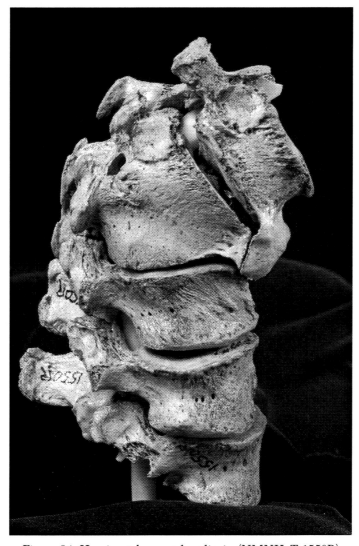

Figure 94. Hemivertebrae and scoliosis. (NMNH–T 1550R)

itation of the vertebral body by erosive degeneration from infection (**Figures 95, 96**). Destruction of the vertebrae usually first begins in the anterior, inferior portion of the vertebral body causing loss of vertebral height, loss of definition of the endplates, collapse and subsequent severe angulation of the spine. Localized destruction in the form of cavitation and abscessing of the vertebral bodies is an inflammatory response to invading tubercle bacilli resulting in the formation of tubercles that stimulate erosion of the trabeculae and cortical bone. Note the minimal new bone formation in comparison to the amount of bone that has been resorbed. Adjacent vertebrae may show destruction where the inflammatory stimulus spreads beneath the anterior longitudinal ligament (sometimes the disease will "skip" vertebrae and

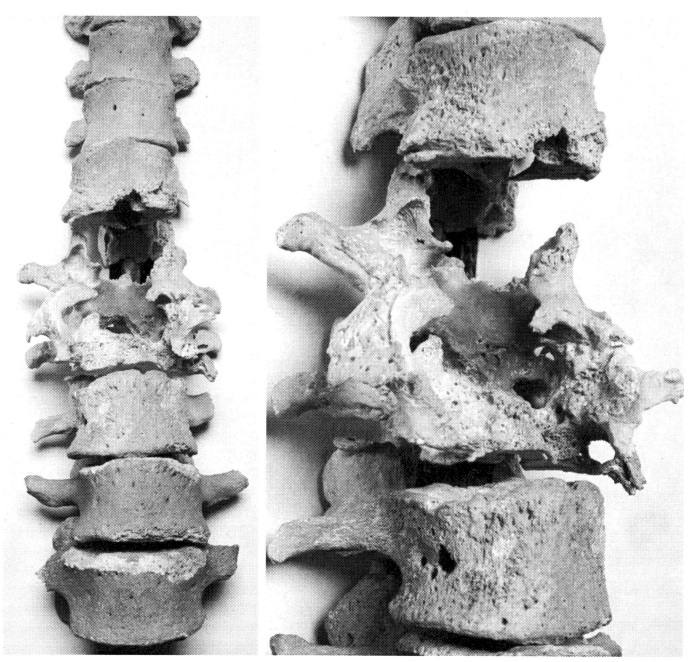

Figures 95a & b & 96. Tuberculosis of the spine (thoracic and lumbar vertebrae), also known as "Pott's" disease. (Fig. 95–NMNH–T 1124R; Fig. 96–AFIP)

continue above or below the initial vertebral abscess). Other frequent sites of tubercular involvement are the inner surfaces of the ilium (psoas abscess through the spread of the disease along the psoas muscle) (see **Fig. 118**), skull, joints (usually only an isolated joint is affected), and ribs. Tuberculous spondylitis occurs in approximately 1 per-

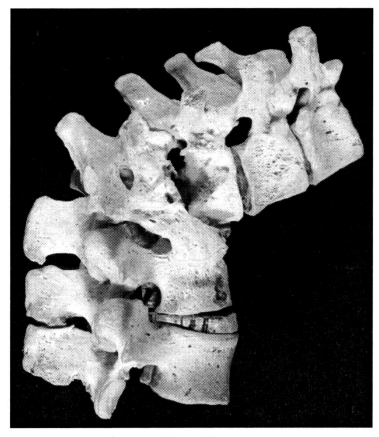

Figure 96.

cent of patients with tuberculosis and is the most common form (25% to 60%) of skeletal tuberculosis (Lindahl et al. 1996; Patankar et al. 2000). Pott's disease (spinal infection) is the most common form of musculoskeletal tuberculosis, associated with nearly 50 percent of all cases (Kramer et al. 2004). Uncommon finding depending on the population under study. Apley and Solomon 1988; Arriaza et al. 1995; Aufderheide and Rodriguez-Martin 1998:121–124; Buikstra and Williams 1991; Fletcher et al. 2003; Ganguili 1963; Garcia-Lechuz et al. 2002; Hallock and Jones 1954; Hodgson et al 1969; Kelley and El-Najjar 1980; Mays et al. 2001; Moore and Rafii 2001; Morse 1961, 1969; Ortner and Putschar 1985; Ortner 2003:230–235; Pálfi et al. 1999; Rom et al. 2004; Steinbock 1976:176–182; Thijn and Steensma 1990 (radiographic study); Tuli 1975; Zimmerman and Kelley 1982:102–108; Zink et al. 2001.

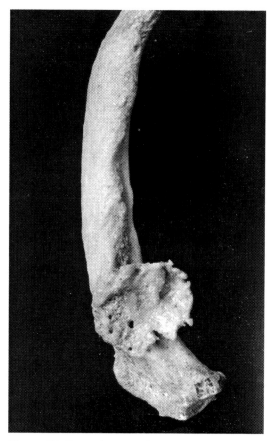

Figure 97. Osteophytes and porosity of the articular facets (OA).

RIB

Raised, irregular bony growths where the rib attaches to the transverse process or facet on the centrum (**Figure 97**). Common finding accompanying middle and old age.

The formation of an irregular "false" joint is produced due to muscle tension and/or when there is continued motion between the two ends of a complete fracture. In some cases when the broken ends fail to properly unite, cartilage develops instead of bone (callus), and a more permanent pseudoarthrosis results (**Figure 98**). Movement of this false joint may properly function for many years after its formation. See also Figure 148a & b in this volume for pseudoarthrosis of the ulna. For a fracture site to properly heal the segments of the bone must be properly realigned and restricted from movement (immobilized in a cast) and muscular tension and motion at or near the site must be minimized (traction). Uncommon to rare finding. Ortner 2003: 151–153; Steinbock 1976:21–22.

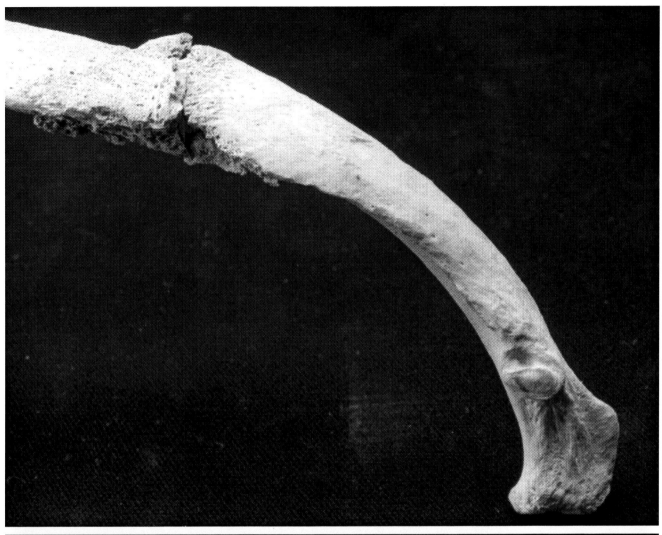

Figure 98a & b. Pseudoarthrosis (pseudarthrosis, malunion, false joint, incomplete union, nonunion). (NMNH Egypt)

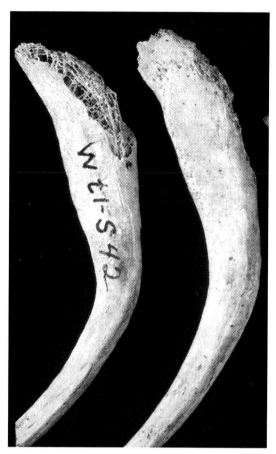

Figure 99. Active or healed periostosis (possibly associated with tuberculosis). (Wt1-S42)

Enlarged and thickened periosteal bone (**Figure 99**) confined to the cortical surfaces of the ribs gives them a clubbed appearance at the vertebral ends especially. Small, roughly ovoid and well-defined areas of reactive bone along the dorsal (pleural/lung side) surface of the ribs may be indicative of tuberculosis (Marc Kelley pers. comm. 1989; Pfeiffer 1991; Roberts et al. 1994, 1996). The etiology of ribs that are greatly thickened down the majority of the shaft is unknown, although tuberculosis and other chronic pulmonary diseases are suspected. Carefully examine the entire skeleton (especially the spine) for other indications of infection or malformation. Uncommon to rare finding in archaeological samples but common in early twentieth-century specimens (pers. comm., Charlotte Roberts 1989). Kelley and Micozzi 1984; Lambert 2002; Roberts et al. 1994, 1996; Santos and Roberts 2001.

These ribs (**Figure 100**) are noticeably straight as a result of chronic kyphosis from severe osteoporosis (osteomalacia) or otherwise dis-

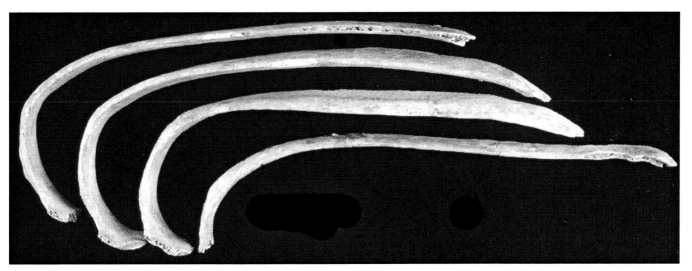

Figure 100. "Shepherd's crook" ribs. (NMNH–T)

eased individuals. The ribs lose their natural "C" shape when the torso is bent forward and causes the internal organs to push against the rib cage, thus straightening the ribs. Nugent et al. 1984; Pitt 1991.

Bicipital ribs (merged ribs) consist of two ribs that fuse to form one (Resnick 2002). Bicipital and bifurcated ribs differ in that the former (**Figure 101b**) are true fusions of what should have been two independent ribs, usually those associated with the first and second thoracic vertebrae (Aufderheide and Rodriguez-Martin 1998:68–69). In this example, the two vertebral ends of the ribs are separate and their bodies are fused along the middle segment. Uncommon to common finding. Not to be confused with congenital fusion of many ribs associated with certain syndromes (e.g., Klippel-Feil 1912). Barnes (1994:71–77) covers the various forms of segmental developmental disturbances which produce bifurcation and bicipital ribs from caudal shift of the cervicothoracic border (Popowsky 1918).

Cervical ribs (**Figure 102**) are enlarged costal elements (the bony bridges that span from the centrum to the rudimentary transverse process of cervical vertebrae) of the seventh, sixth or fifth cervical vertebra (Grant 1972; Yochum and Rowe 1996). Such ribs may be rudimentary or fully developed and roughly similar in shape to normal first ribs but smaller and less "U" shaped. The developmental changes are due to shift in the cervicothoracic border in early developmental growth (Barnes 1994:99–104). Radiologically, cervical ribs are found twice as often in females, 50 to 66 percent occur bilaterally, and in 5.6 per 1000 patients (Grant 1972; Grainger et al. 2001; Yochum and Rowe 1996). While Moore and Persaud (2003) report these ribs as

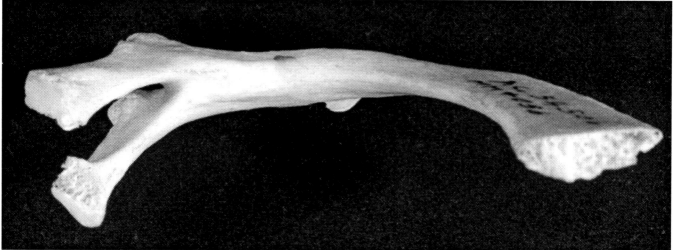

Figure 101a & b. Variations of bicipital and bifid/bifurcated ribs (irregular segmentation of ribs; Barnes 1994). (NMNH–H)

occurring in approximately 0.5 to 1 percent of people, they are uncommon findings in archaeological samples. Erken et al. (2002) found a strong relationship between the presence of cervical ribs and sacralization. Aufderheide and Rodreguez-Martin 1998:68–69, Fig 4.13; Barnes 1994:99–104; Brannon 1963; Cave 1941; Dwight 1887;

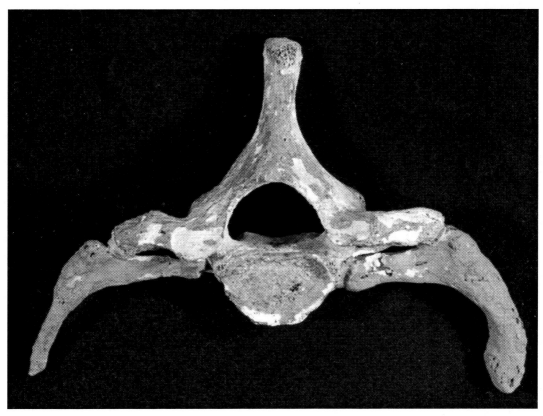

Figure 102. Cervical rib (accessory rib).

Gladstone and Wakeley 1931–32; Hertslet and Keith 1896; Lucas 1915; Magee 2002; Purves and Wedin 1950; Resnick 2002; Schaeffer 1942; Todd 1912.

SACRUM

Sacralization (**Figures 103–106**) is the result of a congenital condition that causes partial or complete fusion of the most inferior lumbar vertebra (fifth or even sixth lumbar) to the sacrum (or when the lowest lumbar vertebra resembles or has features of a first sacral segment (Barnes 1994:108–113; Yochum and Rowe 1996)). This condition has been found in approximately 6 percent of individuals in clinical studies (Grainger et al. 2001). Occasionally a vertebra will exhibit characteristics of a lumbar vertebra on one side and sacral morphology on the other resulting in the so-called hemilumbarization or hemisacralization (Bergman et al. 1988). Although a common clinical finding, these conditions are uncommon in most archaeological samples.

Lumbarization occurs when the first sacral segment has shifted cra-

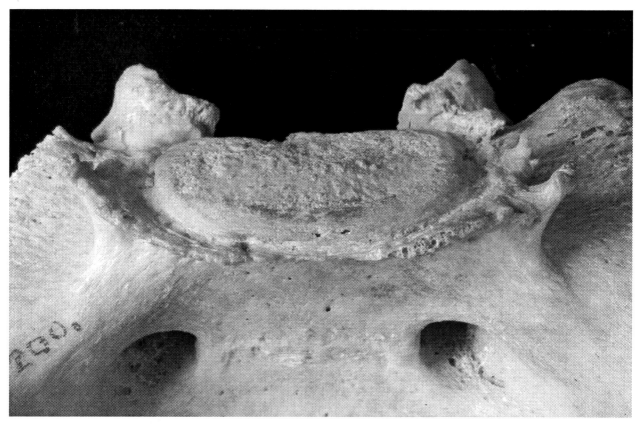

Figure 103. Osteophytes (OA/DJD) along the rim (promontory) of the first sacral segment. (NMNH–T 200) See **Figures 126–127** for fusion of sacrum and ilium.

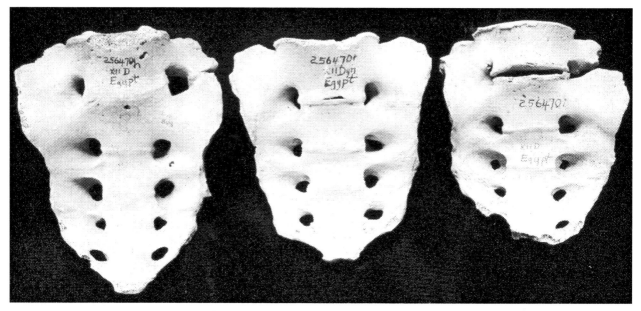

Figures 104, 105 & 106. Sacralization (sacralized lumbar vertebra, transitional lumbosacralization). (Fig. 104–NMNH 256470; Fig. 105–AFIP; Fig. 106–NMNH–256470)

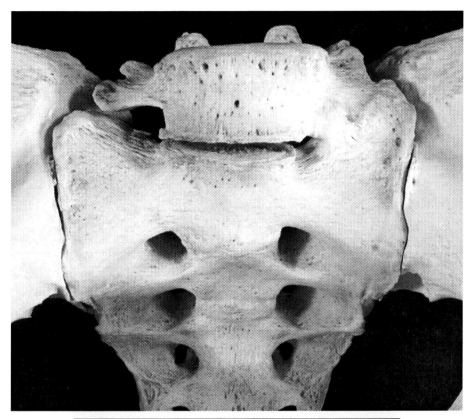

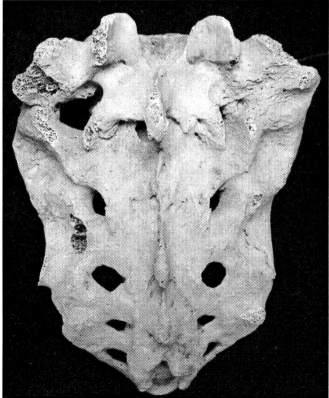

Figures 105 & 106.

nially in the developmental field and has the features of a lumbar vertebra. One way to detect the presence of a transitional, accessory or absent vertebra is to count down the spine beginning with C–2, rather than counting up from L–5. Using this technique, Hahn et al. (1992) found 15 cases of sacralized L–5 and nine cases of lumbarized S–1 in 200 patients. Both sacralization and lumbarization represent transitional vertebrae. Normal five segmented sacra (four sacral foramina) can be distinguished from six segmented sacra (five sacral foramina) by counting the number of paired sacral foramina. While approximately 95 percent of people have 7 cervical, 12 thoracic and 5 lumbar vertebrae, about 3 percent have 1 or 2 more ("extra") vertebrae and about 2 percent have one less (Moore and Persaud 2003). When computing stature, allowance must be made if there is an additional vertebra in the spine (Lundy 1988). Barnes 1994:108–113, 250–255; Cimen and Elden 1999; Hahn et al. 1992; Kim and Suk 1997; Leboeuf et al. 1989.

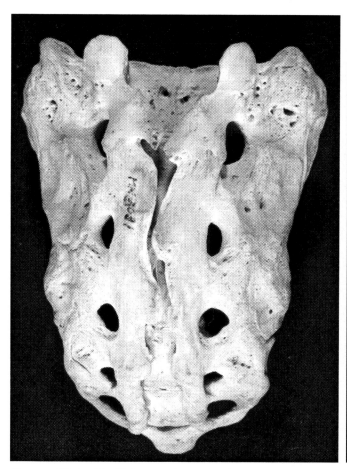
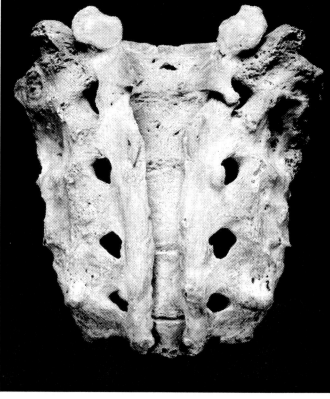

Figures 107 & 108. Cleft spine (spina bifida, cleft spine, incomplete spina bifida)–incomplete closure, fusion, or development of any of the neural arches (spines). (Fig. 107–AFIP MM3031; Fig. 108–NMNH 256470)

This condition may affect any vertebrae, but is most commonly found in the sacrum, especially in L–5 or S–1, occurring in approximately 15–18 percent of individuals with a reported incidence of 9 percent in females and 13 percent in males (Cowell and Cowell 1976) (**Figure 107 & 108**). Two classes are identified, spina bifida occulta and cyctica and having two grades, meningocele and meningomylocele. Occulta class is the more common and least severe of the two. Cyctica is a more rare condition and often results in paralysis due to the spinal cord extruding from the neural arch and traumatized and rates of expression for cystica have been reported by Fishbien (1963) and Norman (1963) at approximately 3/1000. There may be hip deformity and dislocation which will exacerbate the sacral condition (Canale et al. 1992). Besides congenital hip displasia, spina bifida has also been associated with other developmental problems such as hydrocephalus, rib anomalies and clubfoot (Zimmerman and Kelley 1982:31). Uncommon to common finding in populations due to its hereditary properties. Barnes 1994:41–49, 259–265; Webb 1995:235–241; Zimmerman and Kelley 1982:29–31.

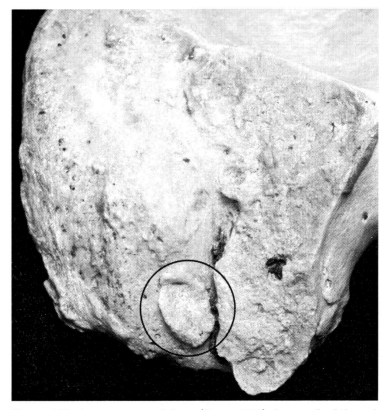

Figure 109. Accessory sacral facet (Derry 1911)–frequently, bilateral accessory facets are present at the level of the first or second dorsal sacral foramen for articulation with the ilium. (NMNH–T)

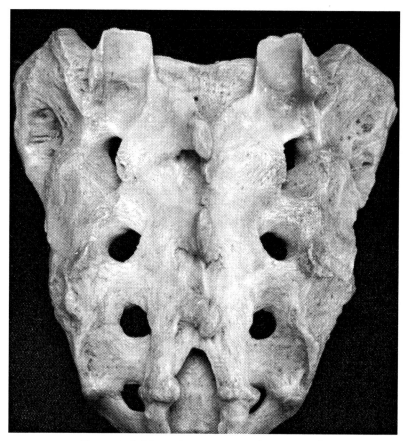

Figure 110. Normal sacral hiatus. (NMNH–T)

The raised contact areas in the ilia are usually located in the most inferior portion posterior to the auricular surface (**Figure 109**). To detect these facets articulate the sacrum and ilium. Common finding.

A canal that extends up one or two segments from the most inferior segment of the sacrum (**Figure 110**). If the hiatus extends as far as the third sacral segment/third sacral foramen (starting with the most inferior segment), it should be scored as spinal bifida. There is no total agreement about the critical criteria among researchers.

INNOMINATE

Macroscopically, enthesophytes (some researchers mistakenly refer to these projections as osteophytes) appear as spike-like projections, spicules, spurs, and ridges or irregular ossification where tendons and ligaments attach (enthesis) to the bone (**Figure 111**). This new bone growth, due to inflammation, may accompany old age, obesity, or repeated acute minor stress related to a particular motion or activity.

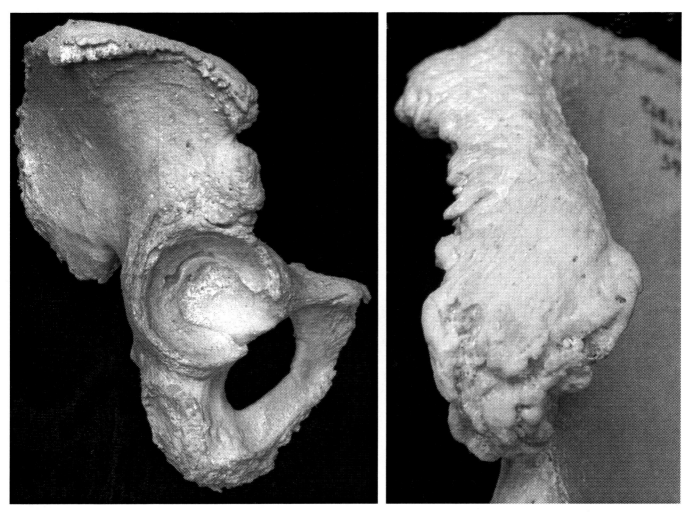

Figure 111a & b. Enthesophytes (Dieppe et al. 1986; Resnick and Niwayama 1988; Taylor and Resnick 2000). Note especially the iliac crest and ischial region. Figure 111b is a close-up of the superior iliac spine. (NMNH 387665)

Frequent sites of involvement are the linea aspera, trochanteric fossa, greater and lesser trochanters of the femur, iliac crest, ischial crest and tuberosity, ischial spine, and obturator foramen in the innominate; attachment of the Achilles tendon in the calcaneus, supinator crest of the ulna, radial tuberosity of the radius, and soleal line (popliteal muscle) of the tibia. In some cases, it is difficult to distinguish enthesopathy from normal skeletal variation (robustness). Enthesophytes, possibly reflecting DISH or fluorosis, are common findings in individuals 60 plus years of age, and may be more prominent on one bone or one side of the body as a result of increased pulling stresses at these sites such as handedness or paralysis. Enthesophytes do not reflect osteoarthritis (inflammation of synovial joints; **Figures 113–115 and 123**) and may accompany leprosy (Carpintero-Benitez et al. 1996).

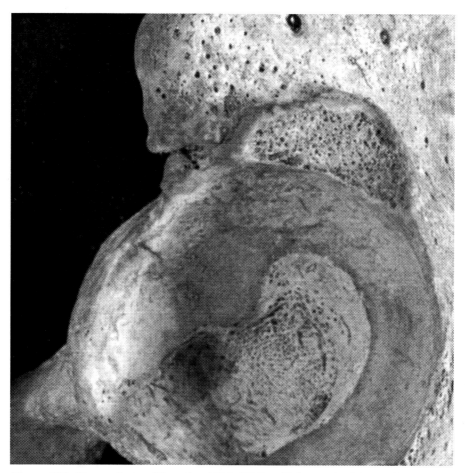

Figure 112. Flange lesion (Knowles 1983: Wells 1976), cotyloid bone (Steele and Bramblett 1988) or os acetabuli/marginal acetabular epiphyses (Resnick 2002; Taylor and Resnick 2000).

Benjamin et al. 2000; Maas et al. 2002. For assessment of entheso-phyte and osteophyte correlations, see Rogers et al. 1997.

The cause and correct name of this condition (**Figure 112**) have not been definitely established. With some reservation, Wells (1976) and Knowles (1983) attribute a similar condition to acute, but temporary, dislocation of the femoral head on the rim of the acetabulum (**Figure 124**). This "lesion," present in one or both innominates, is approximately 2 to 3 cm in length and appears as an eroded area with either exposed trabeculae or as a smooth, flattened depression. (See Hergan et al. 2000; Plotz et al. 2003; Ponseti 1978; Yochum and Rowe 1996 for discussion of os acetabuli.)

Earlier work by Terry (1933) describes a cotyloid bone as a large separate bone in the pubic region of the acetabulum and resembling a bipartition (clearly not the same condition described by Knowles or

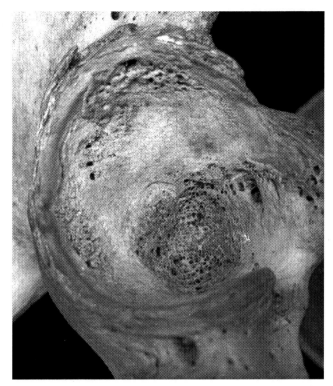

Figure 113. Osteoarthritis (pitting and bony buildup) of the right acetabulum.

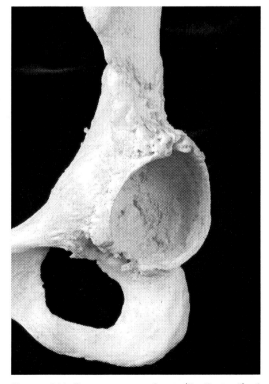

Figure 114. Severe osteoarthritis ("collaring") of the left hip socket. (NMNH)

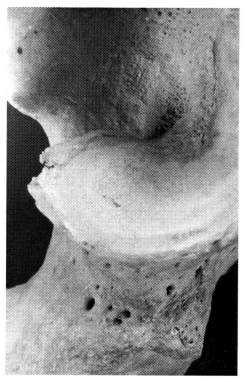

Figure 115. Marginal osteophytes ("lipping") on the left acetabulum reflecting degenerative joint disease (DJD). The porosity in the nonarticular portion of the acetabulum is normal.

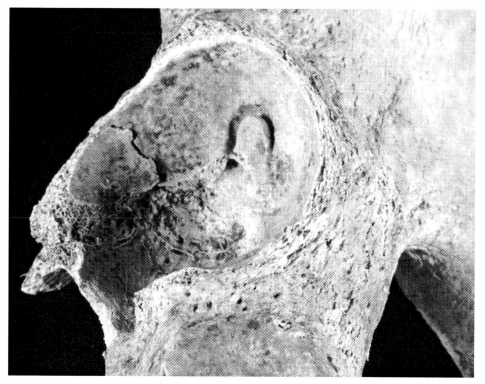

Figure 116. Acetabular mark or notch (normal variant).

Wells). However, there is an example of a cotyloid bone (labeled as such) fitting the description of the flange lesion in the Hamann-Todd collection in Cleveland, Ohio (Marc Kelley pers. comm. 1989). The latter example has the appearance of an accessory bone attached (post-mortem) to the rim of the acetabulum with a wire. Regarding etiology, an accessory bone along the rim of the acetabulum as described as Taylor and Resnick (2000) is more plausible, and similar in appearance to other accessory bones, than traumatic dislocation of the femur or trauma associated with activity as suggested by Stirland (1991b). Either flange lesions, cotyloid bones and os acetabuli represent different conditions (traumatic versus developmental), or terminological confusion obscures the issue.

The authors have only encountered such lesions in two young individuals, both showing the porous form bilaterally most likely associated to bilateral hip displasia (see Aufderheide and Rodriguez-Martin 1998:69–70 for discussion). Care must be taken not to misinterpret normal irregularity in the form of indentations or undulations of the acetabular rim with this condition. Examine the head and neck of the corresponding femur for signs of trauma or alteration reflecting OA. If encountered, it is probably best not to identify the condition by name but rather carefully describe the lesions. If, like the authors believe, Resnick (2002) Taylor and Resnick (2000) are referring to flange lesions, they report that these triangular ossicles adjacent to the superior acetabular rim are present in about 5 percent of the population. Rare finding, however, it is present in most skeletal samples.

A triangular-shaped defect or nearly detached u-shaped "tag" or "ear" of bone located in the superior third of the acetabulum (**Figure 116**). This nonmetric trait may be a remnant of fusion of the bones forming the acetabulum (tri-radiate cartilage). Common finding. Anderson 1963; Saunders 1978.

Enlarged nutrient foramina (**Figure 117**) can be the result of many diseases and conditions, including leprosy, tuberculosis and normal variation. Caution, it is better to note the presence of this condition rather than attempt to "diagnose" it.

A concavity or perforation in the inner surface of the ilium (**Figures 118–120**) usually resulting from neoplasm, tubercular (tuberculosis), actinomycotic or syphilitic invasion via the psoas muscle from the lumbar vertebrae (Simpson and McIntosh 1927). Rare finding in most skeletal populations. However, a skeletal sample with many cases of tuberculosis will show this trait in varying frequencies. Fitoz et al. 2001; Franco-Paredes and Blumberg 2001; Ganguili 1963; Hallock and Jones 1954; Hodgson et al. 1969; Micozzi and Kelley 1985; Morse

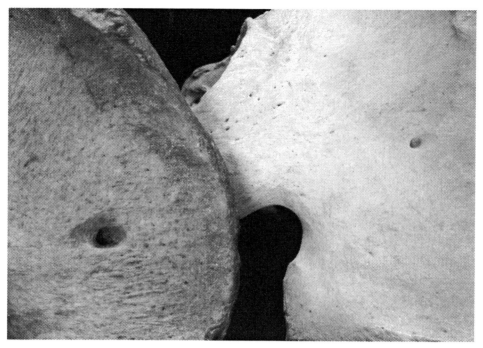

Figure 117. Enlarged nutrient foramen in comparison to normal. (NMNH–T)

1961, 1969; Muckley et al. 2003; Pálfi et al. 1999; Steinbock 1976:176, Fig 71; Versfeld and Solomon 1982; Younes et al. 2002.

Depending on the shape and elevation of the sacroiliac articulation, a groove of varying width and depth will be found anterior to the auricular surface (**Figure 121**). This is generally believed to be a nonmetric sex indicator on the female ilium (Bass 1995:208–218).

The preauricular sulcus (Zaaijer 1866) is often attributed to be the result of pulling stresses of the anterior sacroiliac ligaments during birth delivery. Some researchers, however, distinguish two types of preauricular grooves–the groove of pregnancy and the groove for ligamentous attachment (Houghton 1974, 1975; Saunders 1978); both are located anterior to the auricular surface of the ilium. Some researchers attribute the groove of pregnancy to pulling stresses of the ventral sacroiliac ligament and subsequent inflammation (bleeding) associated with childbirth. Hormonal responses such as relaxin result in a loosening of the sacroiliac joint and localized growth resulting in a larger birth canal. A groove of pregnancy differs from a groove of ligamentous attachment in that the former will generally show discrete or coalesced pits or craters within the groove (which extends more superiorly along the anterior border of the auricular surface of the ilium). The latter will simply be a shallow, short groove present in both sexes (normal imprint of a strong ligament) (Houghton 1975).

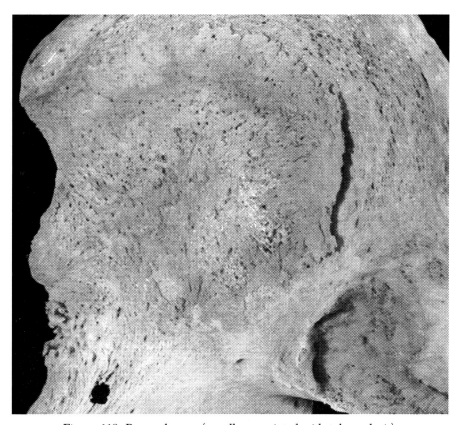

Figure 118. Psoas abscess (usually associated with tuberculosis).

One study by Spring et al. (1989) designed to determine the relationship of the preauricular sulcus and parity (pregnancy) revealed in 190 women a sulcus in 4 of 41 (10%) nulliparous women and 25 of 149 (17%) women with positive pregnancy histories. A review of 3,200 pelvic x-rays (1508 males and 1692 females) by Gulekon and Turgut (2001) revealed preauricular grooves in only 393 of the females (23%). The authors concluded that the presence of a deep preauricular sulcus was not necessarily an indication of past pregnancy. Common finding. Dee 1981; Hatfield 1971; Houghton 1974, 1975; Kelley 1979; Saunders 1978; Spring et al. 1989; Tague 1990 (for discussion of preauricular area in *Macaca mulatta*).

Circular or linear depressions or grooves (**Figure 122**) can be found on the dorsal surface of the pubic symphyses usually in females, but can be found in males at approximately a 4 percent frequency (Suchey et al. 1979, 1986). These pits and/or grooves may resorb in old age. Holt (1978) examined 68 female pubic bones with known medical histories and found that nearly 38 percent of the females who had not given birth had small to large pits and scars on the dorsal pubic bones.

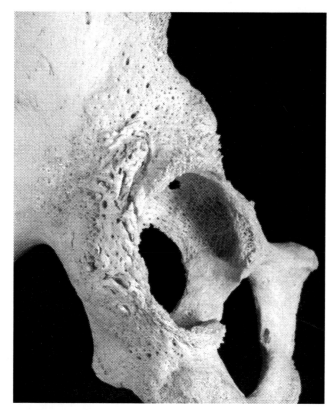

Figure 119. Tubercular destruction of the acetablum with reactive bone around the socket. Ganguili 1963; Versfeld and Solomon 1982; Resnick and Niwayama 1988.

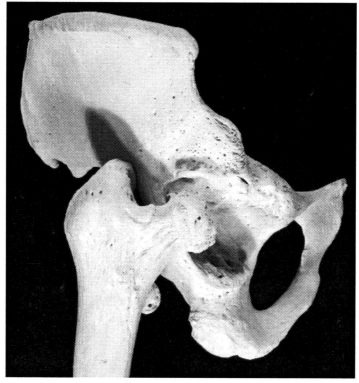

Figure 120. Tubercular destruction of the right femoral head and hip socket (see Apley and Solomon 1988 for several good radiographic examples—for a close-up image of the proximal femur, see Fig. 160).

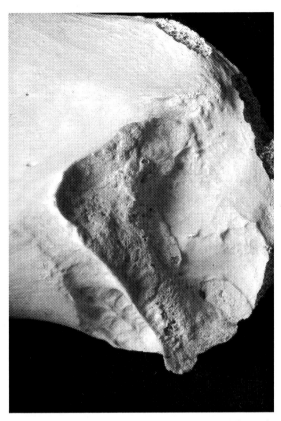

Figure 121. Preauricular groove/preauricular sulcus (groove of pregnancy, parturition groove). (NMNH–T 1173)

Snodgrass and Galloway (2003) examined the pubic bones of 148 modern females and concluded that pressure to soft tissue structures during pregnancy may be a critical factor in the development of dorsal pits. The authors cautioned, however, that further research is needed before drawing any definitive conclusions on the relationship between dorsal pits and number of births. Acute trauma to the pubic symphysis in males from a fall or accident may produce similar dorsal pitting. These findings cast serious doubt on the veracity of using dorsal pitting as an indicator of pregnancy and number of births. Common finding in females. Holt 1978; Kelley 1979; Stewart 1957, 1970; Suchey et al. 1979.

    Posterior superior.
    Posterior inferior.
    Anterior superior.
    Anterior inferior.

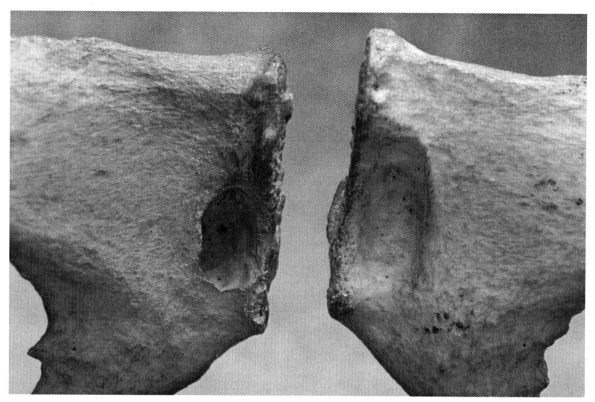

Figure 122. Dorsal pits or pitting (parturition pits, birthing scars).

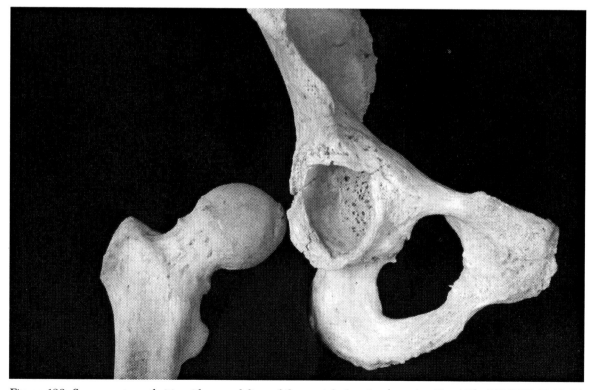

Figure 123. Severe osteoarthritis with remodeling of the acetabular rim (see also Fig. 116). (NMNH–T 321R)

Figure 124. Common positions of the dislocated femur head.

In this condition (**Figure 125**) the urinary bladder may open into the external surface of the abdominal wall (extrophy) resulting in an unusually wide gap (diastasis) between the pubic symphyses. A rare developmental defect that may be accompanied by other skeletal conditions including acetabular dysplasia, an abnormal lumbosacral spine, increased tibial torsion, patellofemoral instability (clinically) and shortening of the pubic rami. Ait-Ameur et al. 2001; Aufderheide and Rodriquez-Martin 1998:70; Kaar et al. 2002; Ortner and Putchar 1985; Sponseller et al. 1995, 2001; Yazici et al. 1999.

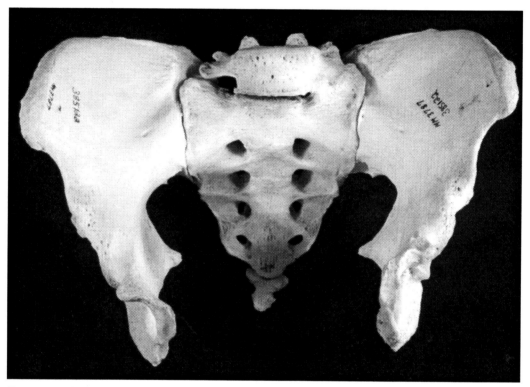

Figure 125. Markedly wide pubic bones associated with extrophy of the urinary bladder (Meschan 1984). (AFIP MM3787)

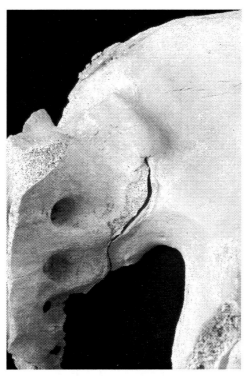

Figure 126. Fused left sacroiliac joint showing a large bridging osteophyte. Regular Sacrolitis may be unilateral or bilateral and is indicative of several conditions, including ankylosing spondylitis or degenerative hypertrophic spondylitis (Ortner 2003:571, Figs. 22–8, 22–10).

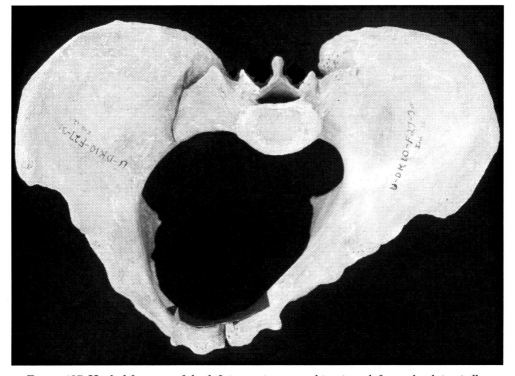

Figure 127. Healed fracture of the left innominate resulting in a deformed pelvic girdle.

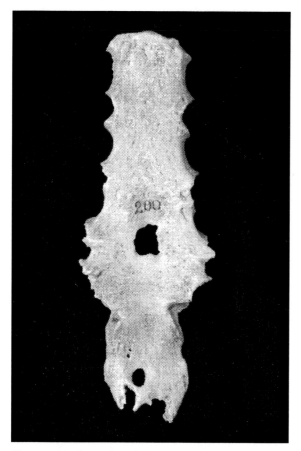

Figure 128. Sternal and xiphoid perforation/foramina/aperture (cleft sternum) (Knight and Morley 1936–37). (NMNH–T 200)

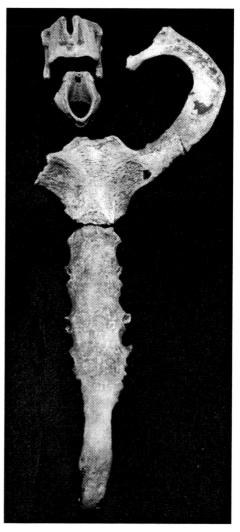

Figure 129. Ossification and elongation of the xiphoid process, and ossification of costal cartilage attaching to the left first rib (also in the photo are the thyroid [Adam's apple] and cricoid cartilage of the throat). All features older age related calcification of the cartilages (see also calcified xyphoid cartilage in Fig. 128). (NMNH 390022)

MANUBRIUM AND STERNUM

This developmental defect is often misidentified as a healed bullet wound or perforating wound of the sternum (**Figures 128-129**). This condition results from the lack of complete fusion of the lower two or three sternal segments, frequently between segments three and four, as they ossify separately from left and right centers (Barnes 1994:227–230; Grant

1972). See Barnes 1994:210–230 for a comprehensive discussion of the multiple variants of morphology in the strnum with numerous citations of the paleopathological literature. Uncommon variant but present in most all populations. Ashley 1956; Cooper et al. 1988; Krogman 1940; McCormick 1981; O'Neal et al. 1998; Resnick 2002; Saunders 1978.

CLAVICLE

The classic rhomboid fossa (**Figures 130–131**) appears as an irregularly shaped crater, groove, depression or excavation along the inferior surface of the clavicle. This fossa associates with the attachment of the costoclavicular ligament (also known as the rhomboid ligament because of its shape) between the sternal end of the clavicle and superior surface of the first rib (Depalma 1963). Other "catchall" varieties of this normal anatomical variant appear as smooth, raised eminences or depressions, crescent shaped ridges and crests. Most generally however, there is no evidence of this trait visible in the clavicle or bony alteration at the attachment site of the costoclavicular ligament. Parsons (1916) noted rhomboid fossae in 10 percent of 183 clavicles. An examination of 10,000 chest fluorograms of individuals between 8 and 70 years revealed this trait in 5 percent of the sample; the youngest individual in the sample with a rhomboid fossa was 11 years old (Khazhinskaia and Ginzburg 1975; Shauffer and Collins 1966).

A preliminary study (contemporary autopsy sample of 350 adults by one of the authors [RWM]) revealed that rhomboid fossae longer than 15 mm in length were usually found in males. Although females commonly showed this trait, it was usually much smaller than those in

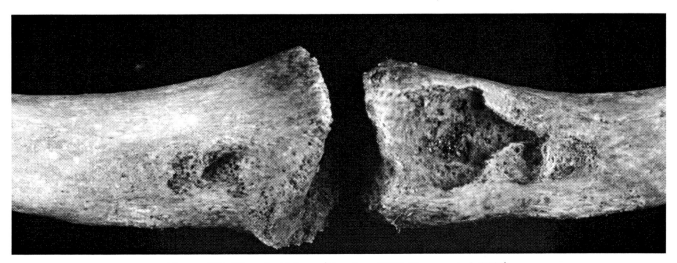

Figure 130. Rhomboid fossa (Depalma 1963; Rogers et al. 2000).

males. Research by Rogers et al. (2000) revealed similar findings, concluding that rhomboid fossae were more commonly found in males and could be used to estimate sex (the largest were in 20–30-year-old males) and that younger individuals rather than older individuals tended to have fossae. Thus, remodeling in older ages reduces the expression of this fossa.

While the etiology of the rhomboid fossa (commonly diagnosed as a lesion on radiographs; Gerscovich et al. 1991) is unknown, it may be aggravated by strenuous activity of the pectoral girdle or dominant hand/arm. The present authors believe that this crater/groove, like those often found in the proximal diaphysis of the humerus (cortical excavations) and the distal femoral cortical excavation (DFCE), is a normal variant that may be especially pronounced in subadults, and is associated with muscle or tendinous insertion. For an excellent overview and discussion of skeletal markers of occupational stress, refer to authors Kennedy, Wilczak, Hawkey, Steen and Lane, Stirland, Robb, Peterson, and Churchill and Morris in the 1998 issue of *International Journal of Osteoarchaeology, 8*(5): 303–411. See Resnick 2002 for more detailed information and radiographic illustrations. Rhomboid fossa is often bilateral, but usually not symmetrical in size and shape. Cave 1961; Hagberg and Wegman 1987; Jit and Kaur 1986;

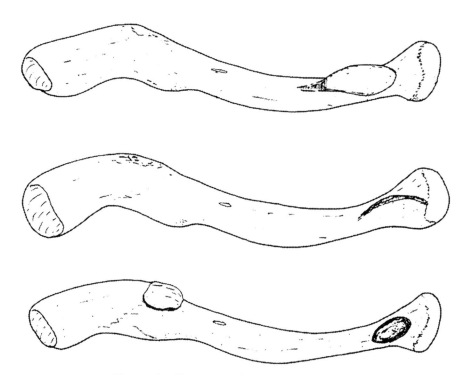

Figure 131. Variations of the rhomboid fossa.

Longia et al. 1982; Rogers et al. 2000; Saunders 1978; Shauffer and Collins 1966; Silverman and Kuhn 1993; Stirland 1991b; Taylor and Resnick 2000; Williams and Warwick 1980.

Upper drawing–Raised plateau-like attachment site for the costoclavicular ligament. Uncommon finding.

Middle drawing–Raised ridge-like attachment site. Uncommon finding.

Bottom drawing–Depressed crater-like fossa–the typical shape is oval or oblong. The fossa will show sharply defined cortical margins and a porous center of exposed trabecula. Common finding.

Conoid joint/process (Cockshott 1958), coracoclavicular joint or bar (Depalma 1963; Taylor and Resnick 2000) or conoid tubercle of clavicle (Yochum and Rowe 1996) on the bottom illustration is a rare finding in most skeletal populations (**Figures 131 & 132**). This raised plateau-like bony extension is a normal variant with no clinical significance

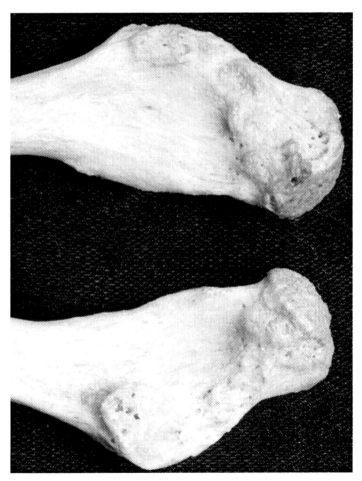

Figure 132. Conoid joint of the clavicle (see conoid joint of the scapula, Figure 137).

that articulates with the superior surface of the coracoid process of the scapula with a reported incidence of 1 to 1.2 percent of individuals. Look for a corresponding facet on the coracoid (**Figure 137**). This is most oftenly expressed bilaterally.

SCAPULA

A nonmetric trait (**Figure 133**) representing an ununited ossification center of the lateral end of the acromion. There are three separate centers of ossification for the acromion process, which will coalesce and reach normal final fusion between 16 and 25 years (Mudge et al. 1984; Park et al. 1994: Rockwood and Matsen 1998). In a radiographic study of 1800 shoulders, Liberson (1937) found the trait bilateral in 62 percent of cases and an incidence of 1.4 percent. Mudge et al. (1984) and Park et al. (1994) further reported that a correlation exists between rotator cuff tears and os acromiale. Sammarco (2000) observed from 1198 individuals from the Hamman-Todd Collection (Cleveland) that 8 percent of the population showed os acromiale with 33 percent being bilateral. Sammarco also found that os acromiale was more prevalent in Blacks (13.2%) while it was much less frequent in Whites (5.8%) and more frequent in males (8.5%) than females (4.9%). A complete survey of the Terry Collection by one of the authors (DRH) resulted in the overall incidence of os acromiale at 8.3 percent and group frequencies ranking greatest in Black males (12.5%), then Black females (9.2%), white males (6.8%) and the lowest occurrence in White females (3.2%) (the combined sex frequencies, 11% for Blacks and 5% for Whites). The authors would derive from these results that there is some genetic component to the occurance of os acromiale, with the significant difference of expression between American Blacks and Whites. Angel et al. (1987) reported a high incidence of os acromiale in the First African Baptist Church cemetery in Philadelphia (8.3%) and patterned special distribution within the cemetery, which they suggest as hereditary inheritence. In contrast, Stirland (1984) found a high frequency (12.5%) of os acromiale in the crew of the tudor ship *Mary Rose* and hypothesized that the condition was the result of pulling stresses on the unfused final element of the acromion related to long-term use of heavy longbows (1991). However, Stirland's sample was composed primarily of young men below the age of 25 years and only the continued unfused epiphyses in individuals over 25 years of age should be scored as os acromiale (Saunders 1978). Angel et al. 1987; Chung and Nissenbaum 1975; Edelson et al. 1993; McKern and Stewart 1957; Resnick 2002; Saunders 1978; Stirland 1984, 1991b; Symington 1900.

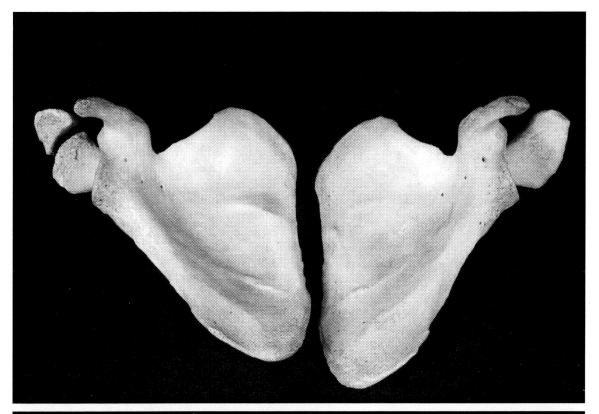

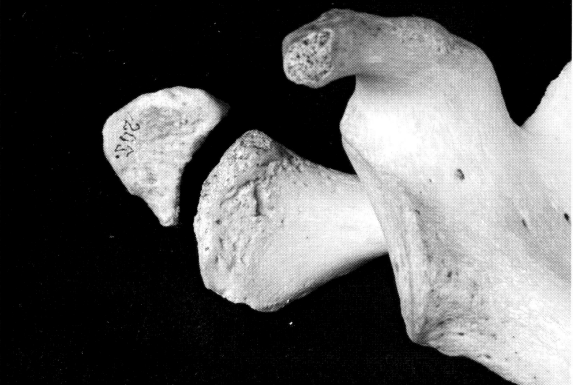

Figure 133a & b. Os acromiale (unfused acromial epiphysis, bipartite acromion, meta-acromion). (NMNH–T 205)

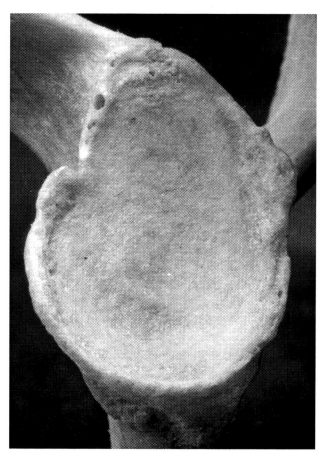

Figure 134. Marginal osteophytes (OA) of the glenoid fossa. (NMNH–T)

Raised, sharp margins along the glenoid fossa (or any other bony joint) can be detected either visually or by using a fingernail (**Figures 134–135**). Common finding accompanying middle age and older as well as individuals functionally stressing their shoulders in various activities (OA). Graves 1922.

Small to large pits (macroporosity) in the articular surface of the glenoid fossa. The normal fossa should be smooth and slightly concave or have an undulating surface. Common finding.

This condition (subluxation/luxation) may result in flattening, erosion, porosity, and eburnation of the glenoid fossa (scapular deformity) (**Figure 136**). Although the condition of dislocation (clinically) may be fairly common in many populations, it is uncommon to rare to find discernible bony changes related to dislocation in either the humeral head or glenoid fossa. Edelson 1993; Ortner 2003:160, Figs. 8–62 & 8–63 & 8–66 & 8–67; Park et al. 1994; Resnick and Niwayama 1988; Steinbock 1967:40.

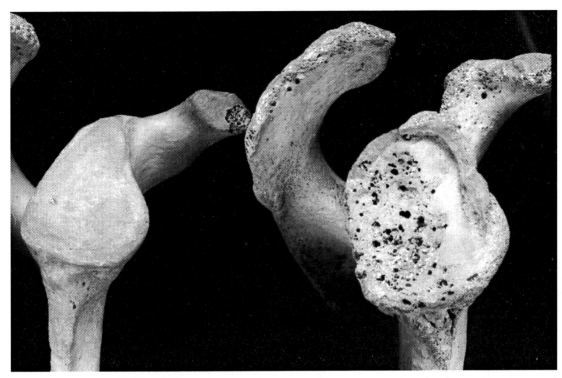

Figure 135. Surface porosity and moderate-severe degenerative joint disease (OA) on right, compared to normal glenoid on left. (NMNH–H)

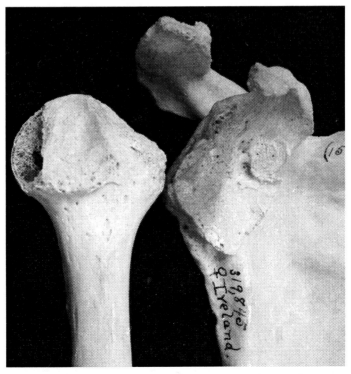

Figure 136. Chronic unreduced dislocation (anterior) of the humeral head (glenohumeral joint). (NMNH–H 319845)

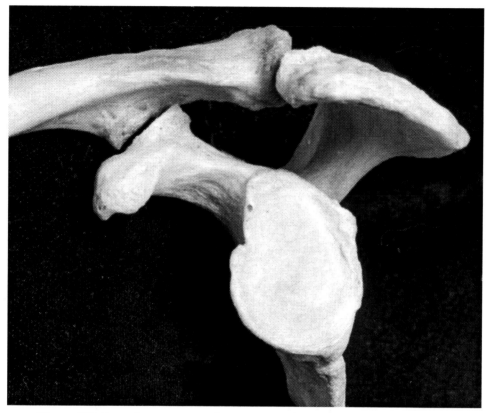

Figure 137. Conoid joint or costoclavicular joint (DePalma 1963). (NMNH–T)

In some individuals the inferior surface of the distal clavicle (**Figuress 131, 132 & 137**) will exhibit a raised, circular plateau/table of bone for articulation with the superior portion of the coracoid process of the scapula. Uncommon finding.

HUMERUS

A porous groove (cortical defect; Brower 1977; Caffey 1972; Keats 1973) upper humeral notches (Taylor and Resnick 2000) for attachment of the pectoralis major muscle (**Figure 138**). In subadults this groove is probably a normal anatomical variant that remodels and later fills in, often leaving a shallow depression in the adult. The pectoralis groove is lateral to the smaller teres groove and both are oriented parallel to the long axis of the diaphysis. Cortical defects are common in children, adolescents (10–16 years of age) and young adults, especially individuals who are unusually physically active, but rare in adults (Mann and Murphy 1989; Murphy and Mann 1989; Ann Stirland Pers. Comm. 1989). (See Cervical Fossa of Allen for a

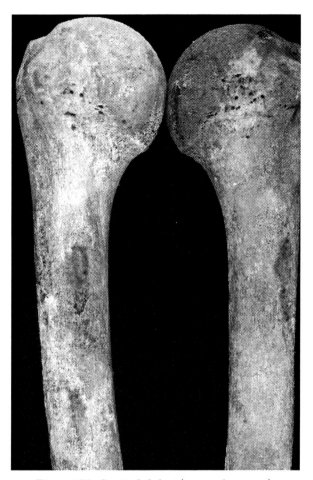

Figure 138. Cortical defect (pectoralis major).

possible relationship between this and cortical excavations elsewhere in the skeleton.) Yochum and Rowe 1996.

The teres major cortical defect is a porous groove similar in appearance to the pectoralis defect noted above. This groove is a normal feature frequently seen in adolescents and young that normally remodels and fills in by adulthood, but may be accentuated (deepened) due to acute or chronic trauma to the shoulder (e.g., prolonged tennis activity). Uncommon to rare finding in adults. Mann and Murphy 1989; Ann Stirland Pers. Comm. 1989. For an excellent overview and discussion of skeletal markers of occupational stress, refer to authors Kennedy, Wilczak, Hawkey, Steen and Lane, Stirland, Robb, Peterson, and Churchill and Morris in the 1998 issue of *International Journal of Osteoarchaeology*, *8*(5): 303–411.

A small, roughly triangular and hook-shaped exostosis (**Figure 139**) projecting 5 to 7 centimeters above the medial epicondyle and varying in length from 2 to 20 millimeters (Genner 1959, Fowler et al. 1959).

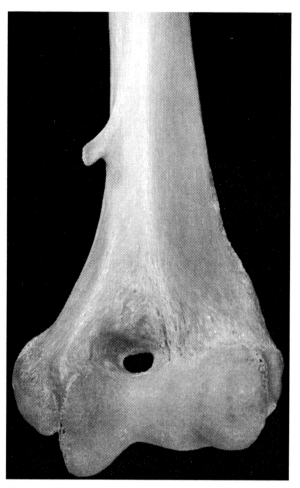

Figure 139. Supratrochlear spur. Supra-condyloid process (Struthers 1873), supracondyloid process (Schaeffer 1942; Fowler et al. 1959), supratrochlear spur (Sunders 1978), supracondylar spur (Kessel and Rang 1966), supracondylar process (Yochum and Rowe 1996). (NMNH–H)

This exostosis serves as an accessory ligamentous attachment for origin of the pronator teres muscle. Through the tunnel formed by this fibrous band (Struther's ligament) passes the median nerve and brachial artery and may result in neurovascular impingement. Congenital trait reportedly found in 7 of 1000 living subjects by Schaeffer (1942) and in approximately 1 percent of people of European ancestry (Barnard and McCoy 1946; Gray 1948; Terry 1921, 1926, 1930). This trait has a high rate of heritability and has been found in embryos (Adams 1934), children of all ages, and adults. Uncommon finding in most archaeological samples. Cady 1921; Dwight 1904; Hrdlicka 1923; Marquis et al. 1957; Parkinson 1954; Rau and Sivasubrahmanian 1931; Witt 1950.

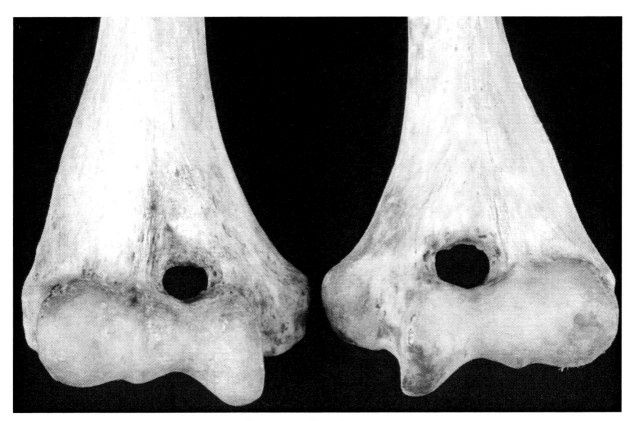

Figure 140. Septal aperture.

A hole present in the olecranon fossa that may range from the size of a pinpoint to the width of a pencil (**Figures 139–140**). Although the etiology of this perforation is uncertain (e.g., congenital, developmental/mechanical or hereditary), it is often found more frequently in females than males and occurs in 4 to 13 percent of individuals (Bergman et al. 1988). Common finding. Bass 1995:54; Hrdlicka 1932; Trotter 1934.

Severe OA of the posterior distal left humerus (**Figure 141**). Note the large osteophytes along the medial margin of the articular surfaces ("mushrooming") and extension into the olecranon fossa. Severe forms of OA of the distal humerus may also result in bony deposits (hyperplasia resembling small mounds), large pits, and distortion of the articular surface. Although similar severe changes may be the result of local acute trauma, most instances are associated with old age. Look for fractures of the proximal ulna and distal humerus. The severe form is uncommon.

Porosity may be interspersed in or along this thin, raised ridge of new bone formed between the trochlea and capitulum, responding to inflammation at the radio-ulnar articulation (**Figures 142 & 143**). Common finding. Ortner 1968.

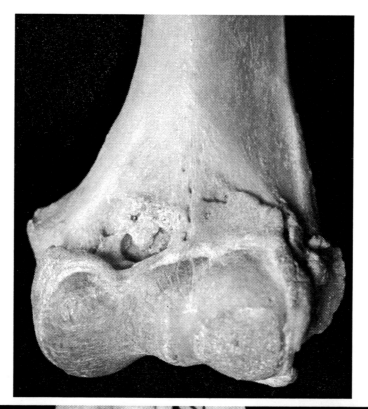

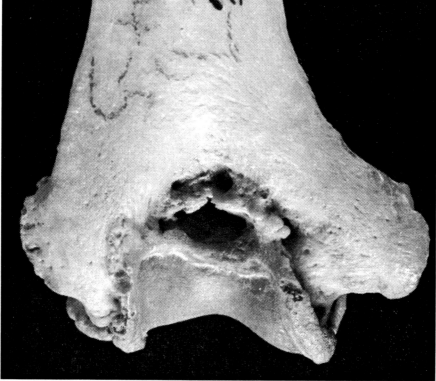

Figure 141a & b. Surface osteophytes of the dorsal surface of the distal humerus and moderate-severe DJD with distortion of the articular margin and eburnation of the capitulum. (NMNH–H)

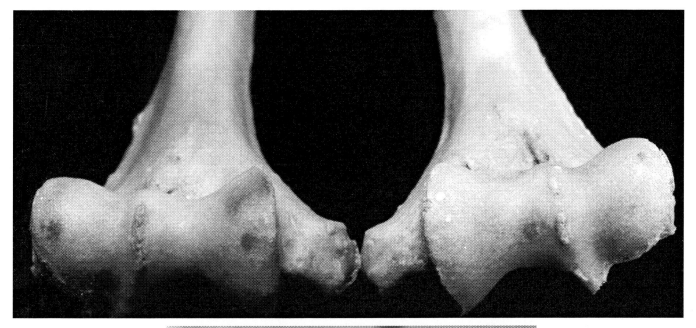

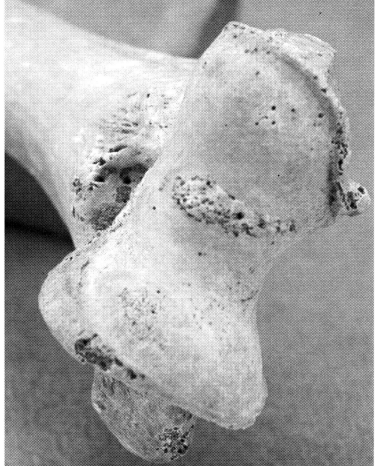

Figures 142 & 143. Early stage osteoarthritis of the distal humerus with a ridge-like buildup of bone.

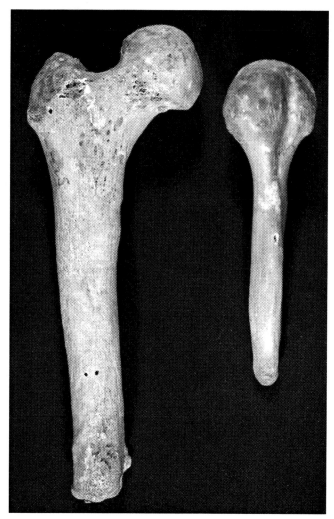

Figure 144. Healed amputated humerus (and femur) at about midshaft.

Evidence of an amputation (**Figure 144**) is apparent in that the normally open/hollow medullary cavity at the site of amputation has closed/remodeled and the distal end of the diaphysis is blunt and rounded (Zimmerman and Kelley 1982:54, Fig. 37). Aufderheide and Rodreguez-Martin 1998:29–30; Steinbock 1976. For references on archeological specimens see Brothwell and Sandison 1967:640–641.

RADIUS AND ULNA

Pitting within and enthesophytic development (**Figures 145 & 147**) along the radial tubercle (especially along the lateral border) is the result of inflammation of the biceps muscle insertion from strenuous

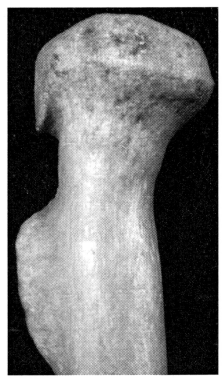

Figure 145. Marginal osteophytes (osteoarthritis) of the proximal radius.

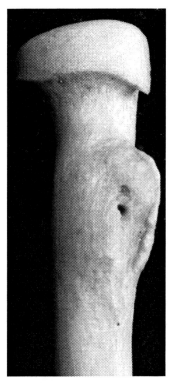

Figure 147. Enthesophyte and pitting of the radial tuberosity.

Figure 146. Pitting form (moderate to severe) of osteoarthritis of the proximal radius. Note the crescent shaped defect reflecting complete loss of cartilage and bony destruction and eburnation due to contact with the distal humerus.

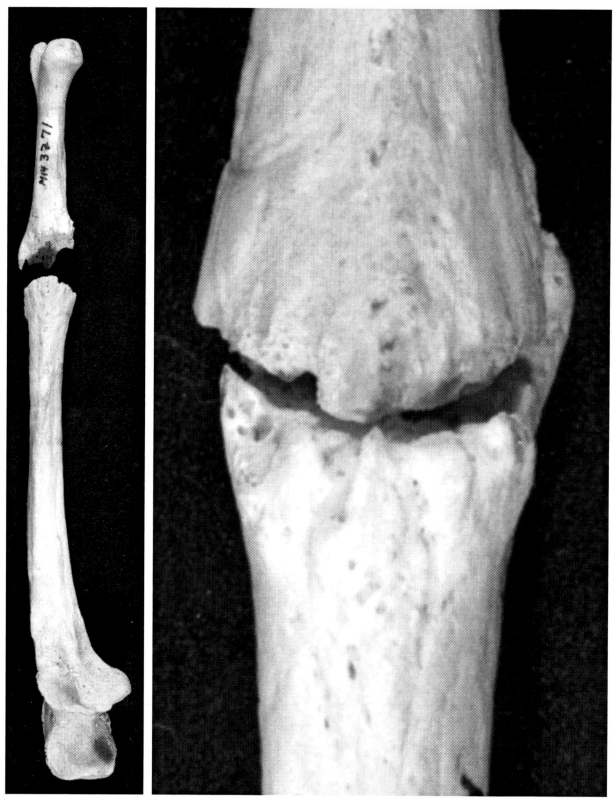

Figure 148. Pseudoarthrosis of the ulna (see also Figure 98a & b for other example of pseudoarthrosis in a rib). (AFIP MM3271)

activity. This feature is a common finding in all populations and will increase in size and severity with age of the proximal radius (see **Figure 146**).

Continued motion between two broken ends of a complete fracture from unrestricted motion can unoftenly produce a pseudoarthrosis (**Figure 148**). Restriction of motion at the site should allow callus development to reunite the bone, unless a cartilage barrier has been formed, resulting in a permanent pseudoarthrosis. See also **Figure 98** for pseudoarthrosis of a rib. Uncommon to rare finding. Ortner 2003:151–153; Steinbock 1976:21–22.

Parry fracture (**Figure 149**) is generally associated with trauma when a person raises their arm to block a blow to the face or head, the ulna taking the brunt of the blow resulting in fracture. Fracture at the distal end of the ulna may also be associated with trauma induced in breaking a fall. Check distal radius for possible associating fracture (Colle's) before assuming this is a defense fracture (Nielsen et al. 2001).

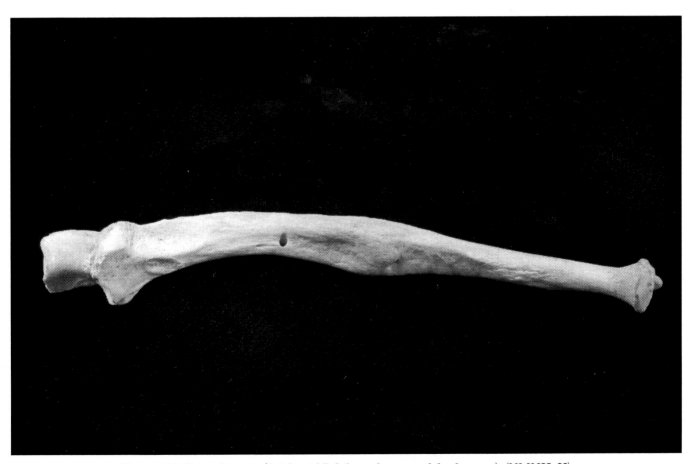

Figure 149. Parry fracture ("nightstick" defense fracture of the forearm). (NMNH–H)

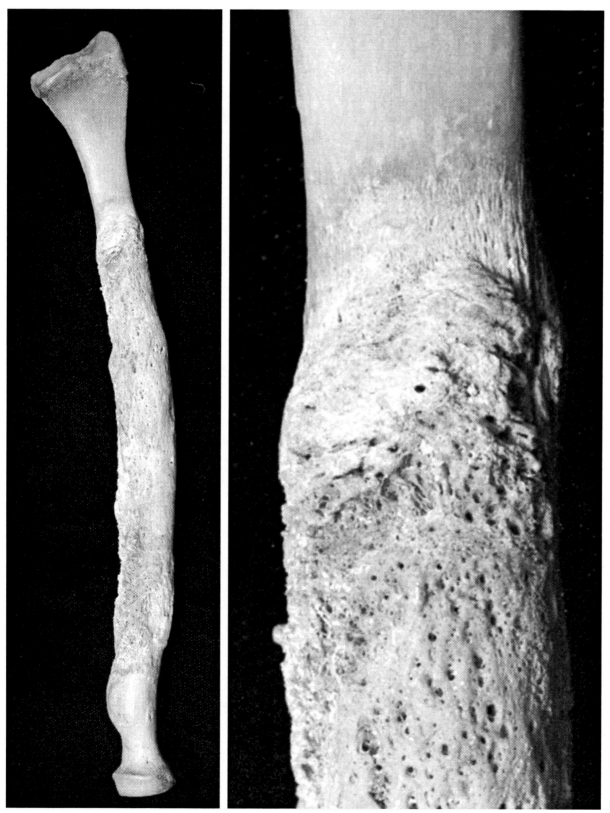

Figure 150a & b. Acute infection (active periostosis) of the radius showing line of demarcation between reactive new bone (diseased) and normal bone. (See Aufderheide and Rodriguez-Martin (1998:179–181) and Ortner (2003:206–214) for other periostosis examples.) (See also **Figs. 183 & 184** for periostosis on tibia.) (NMNH–H 319963)

Interpersonal violence is not always the cause of a Parry fracture, this is emphasized by Grauer and Roberts (1996), and can be the result of accidental and osteoporotic complications (Krishnan 2002; Lindau et al. 1999). Incidence is generally uncommon but is dependant on the amount of warfare or domestic violence in a population. Knowles 1983; Ortner 2003:137–138. See **Figure 150** for example of periostitis of radius.

Erosion of the radial styloid process is one of the criteria associated with rheumatoid arthritis (Resnick and Niwayama 1988). However, osteoarthritis can also deteriorate the styloid process and may resemble a rheumatoid condition. Consult a rheumatologist to evaluate erosion with little or no bony proliferation around the distal ulnar and

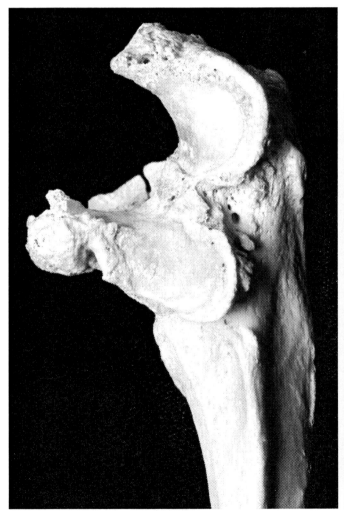

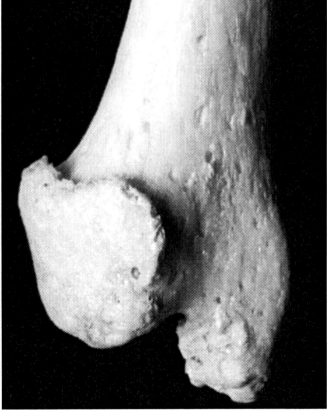

Figure 151. Severe osteoarthritis of the proximal ulna. (NMNH-H)

Figure 152. Osteoarthritis of the distal ulna (marginal osteophytes)–note alteration and blunting of the styloid process. (NMNH-H)

radial articulation (Leak et al. 2003). Fracture of the styloid process is also common in Colle's fracture. Nielsen et al. 2001; Dieppe 1986. Figures 151 and 152 represent arthritic changes of the proximal ulna.

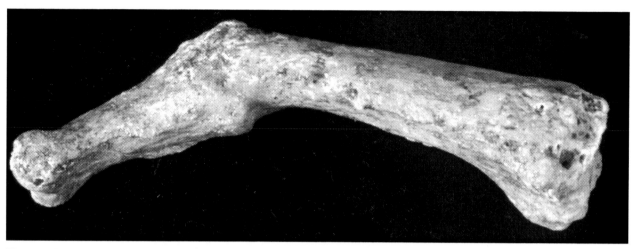

Figure 153. Boutonniere deformity (destruction and, possibly, bony fusion of the proximal interphalangeal joint [PIP]) of the hand.

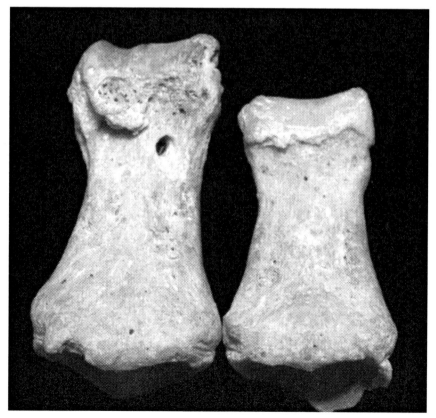

Figure 154. Enlarged nutrient foramen (phalanx). (NMNH–H) Also see **Figure 117**.

HAND

Although this condition is often associated with rheumatoid arthritis, localized trauma and infection must be considered as part of the differential diagnosis when only one joint is involved. Boutonniere or buttonhole deformity (**Figure 153**) results from flexion at the proximal interphalangeal joint and hyperextension at the distal interphalangeal joints. Other skeletal indicators of rheumatoid arthritis include periarticular cysts, loss of joint space, and angulation of joints. Boyer and Gelberman 1999; Coons and Green 1995; Resnick and Niwayama 1988.

This condition, possibly the result of increased vascularity or velocity, can result from a variety of diseases and conditions, including leprosy, but can also be a normal variant (**Figure 154**). In this particular specimen, the enlargement is most likely associated with osteoarthritis as is evident by extension of the distal margin.

FEMUR

Osteoarthritis commonly affects the knee and, the distal femur in the form of marginal osteophytes bordering the articular surface of the

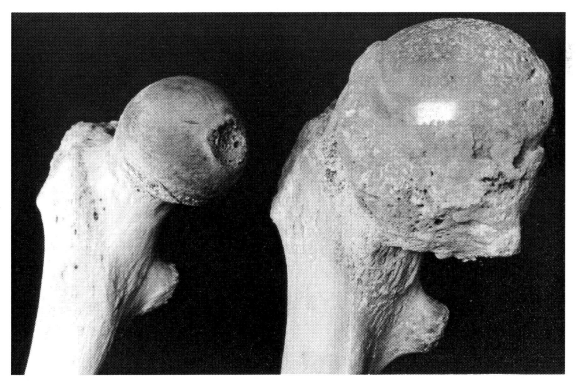

Figure 155. Severe OA/DJD and eburnation (shiny area) of the femoral head compared with normal (left). (NMNH)

condyles and, to a lesser degree, pitting of the articular surface. In severe cases, ivory-like eburnation with flattened and grooved areas may be produced in the articular surface from loss of the cartilage and subsequent 'bone on bone' wear. Eburnated bone can range in size from extremely small (early stage) to areas involving the entire articular surface. Eburnation has a smooth/polished shiny appearance that reflects light and is frequently yellow, resembling old piano keys. Uncommon to rare finding associated with the elderly. See Rogers and Waldron 1995:38, Fig 4.4 & p. 44, Fig. 4–10 for examples. It is also good to perform radiographic investigation to evaluate extent of lipping and cystic changes in the femoral head from necrosis.

Note the followng bony changes:

Periarticular bone (hypertrophic bone, periosteal osteophytes, "mushrooming"). Loss of bone height in the femur head has resulted in new bone growth near and encircling the neck (OA) (**Figure 155**). Most believe that osteoarthritis is the result of changes in the abnormal mechanical forces in and around the affected joint.

Flattened areas of bone where there has been degeneration and resorption of most of the femur head (OA).

Bone cyst (subchondral cyst) below the articular surface. Common radiographic finding in elderly individuals with osteoarthritis and avascular/ischemic necrosis (dead bone usually resulting from trauma and loss of blood supply to the femur head; especially common in very old individuals (**Figure 161**). Claffey 1960; Dieppe et al. 1986; Jaffe 1969, 1975.

If degenerated femoral heads and deformed acetabula are found in children, this pathological condition may be the result of Legg-Calve-Perthes disease (Caterall 1982; Karpinski et al. 1986; Schoenecker 1986; Schwartz 1986) or osteochondritis of the femoral head or slipped capital femoral epiphysis (Busch and Morrissay 1987; Schoenecker 1985; Schiesinger and Waugh 1987) (Schoenecker 1986) but histological investigation should be considered to confirm the bone cell changes which are associated with this disease.

Stirland (1991b), in examining the 179 crewmen of the British ship *Mary Rose* that sank in AD 1545, reported finding six femora (2 left and four right) with "unusual pits" (OD) in the femoral heads, superior to the fovea capitis (**Figure 156**). Aufderheide and Rodreguez-Martin 1998:81–83.

Osteochrontritis dissecans (OD) is also seen as a scooped-out lesion, usually in the medial condyle of the distal femur. First described in 1870 by James Paget, OD is usually attributed to avascular necrosis that begins with acute trauma to the joint (not to be confused with a periarticular cyst that, as the name implies, is located along the articular margin, not on it; **Figure 157**). However, numerous other causes

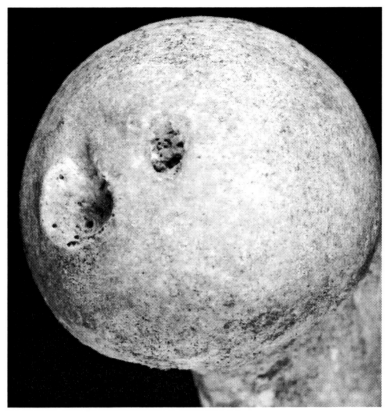

Figure 156. Osteochondritis dissecans (osteochondritic pit) in the femoral head.

have been suggested including abnormal ossification of the epiphyseal cartilage, contact with the tibial spines, hereditary influences, and generalized disorders (Smillie 1960; Wells 1964, 1974).

Most researchers believe the lesion begins with the separation of a portion of cartilage and underlying bone fragment resulting in a "loose body" in the joint ("joint mice") (**Figure 206**). In time the bone and cartilage may resorb, creating a crater-like defect in the mature skeleton (Bradley and Dandy 1989). OD can affect any joint but has a predilection for the medial condyle of the femur and appears during the second decade of life. Hughston et al. (1984) found 78 defects in the medial femoral condyle and 17 defects in the lateral femoral condyle in a sample of 83 patients with OD. Clinically, OD is seen in 15 to 21 cases per 100,000 in the femur although many cases may go undetected unless accompanied by pain and detected on x-ray. Uncommon to common finding in most skeletal samples. Aufderheide and Rodriguez-Martin 1998:81–83, Fig 5.5; Barrie 1987; Clanton and DeLee 1982; DeSmet et al. 1997; Griffiths 1981; Ortner 2003:353, Fig. 13.8; Stirland 1991b, Manchester 1983; Zimmerman and Kelley 1982:72, Fig. 5.5.

Bradley and Dandy (1989:518), in contrast, examined lesions of the femoral condyles in 5,000 knee arthroscopies and found that OD develops as an expanding concentric lesion that forms in an otherwise normal epiphysis. They reported that OD develops in the second decade of life and progresses to a steep-sided defect in the mature skeleton. They further hypothesized that OD and acute osteochondral fractures with or without loose bodies, were different conditions and should not be classified or referred to as OD.

A scooped-out concavity, pit or depression situated along the articular margin of a true joint. If the concavity "cyst" is located on the articular surface, especially near the middle of the joint of the distal femur or proximal tibia, an osteochondritic pit should be considered in the differential diagnosis. Similar pits may result from a variety of reasons including acute or repeated trauma to a joint and infection. Hill et al. 2003; Resnick 2002; Resnick and Niwayama 1988.

Features of this specimen (**Figures 158 & 159**):

Surface osteophytes. Raised, isolated bony growths on the articular surface.

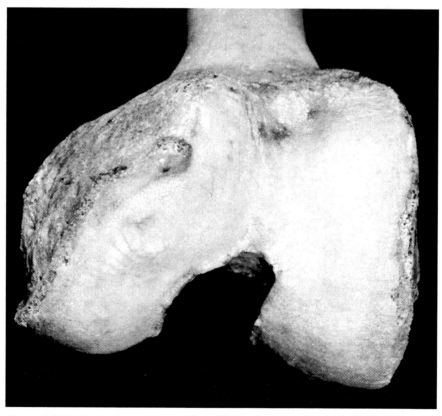

Figure 157. Periarticular cyst of the distal femur (similar cysts may be found at any joint (sometimes referred to as synovial cysts or pseudocysts of osteoarthritis).

Marginal osteophytes. Raised lip of bone along the junction of the head and neck of the femur or fovea capitis (epiarticular osteophytes (Jeffrey 1975)).

Surface porosity. Small to large pits in the articular surface.

Enthesophytes. New bone at the insertions of tendon and ligaments. Enthesophytes appear as irregularly raised areas (excrescences), roughened attachment sites, bony spicules or "whiskers," spurs or projections commonly found on the greater and lesser trochanters, trochanteric fossa, and linea aspera. This condition is due to stresses (inflammation) at tendon and ligament attachments (any) to bone. Some researchers state that enthesophytes occur only at the site of tendon and ligament insertions (Dutour 1986) because more stress is placed on the insertions, rather than the origins, due to the smaller area of attachment fibers into the bone. Enthesophytes are strongly

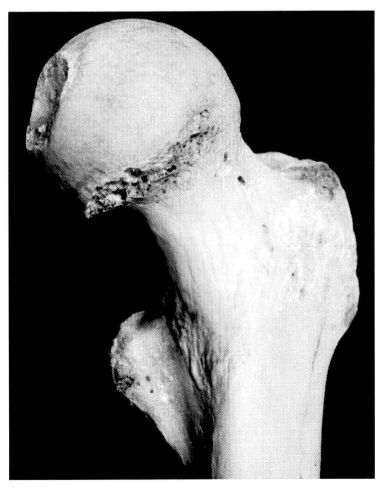

Figure 158. Osteoarthritis and enthesopathy of the proximal femur (anterior view).

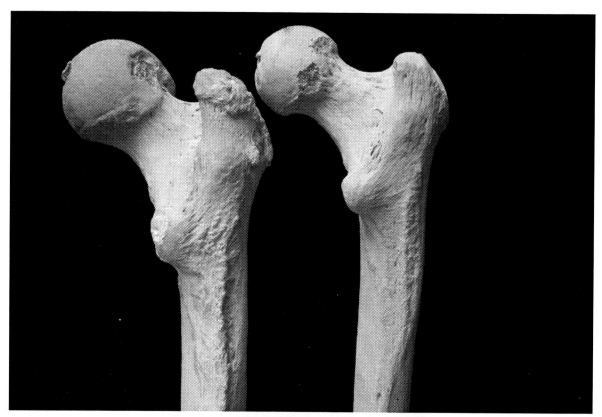

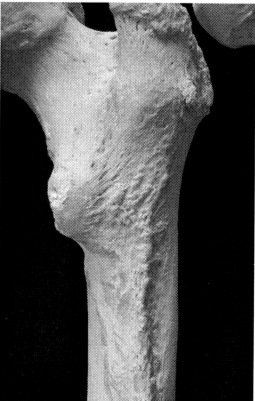

Figure 159a & b. Enthesophytes along the linea aspera.

correlated with old age. Care must be exercised not to confuse enthe-sophytes with myositis ossificans. Dutour 1986; Merbs 1983; Pecina and Bojanic 2004; Resnick and Niwayama 1988.

Irregular spikes and crests of bone at muscle attachment sites along the linea aspera. Enthesophytes, also sometimes known as traction spurs, reflect pulling stresses and inflammation at ligament and tendon insertions.

Surface and marginal osteophytes are common findings in the elderly.

Subchondral bone cysts. X-rays will reveal radiolucent (dark) areas indicating fluid-filled cysts/pockets beneath the articular cartilage (**Figure 161**). Common finding associated with degenerative articular changes accompanying osteoarthritis, rheumatoid arthritis, and aseptic necrosis. The cyctic changes shown can reflect acute or long-term

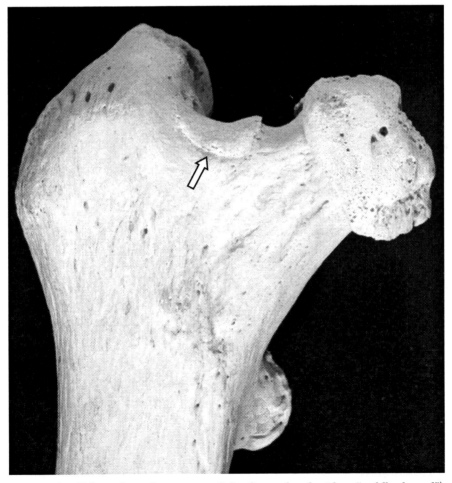

Figure 160. Tuberculous destruction of the femur head with a "saddle-shaped") plaque formation on the neck from articulation with the superior border of the acetabulum.

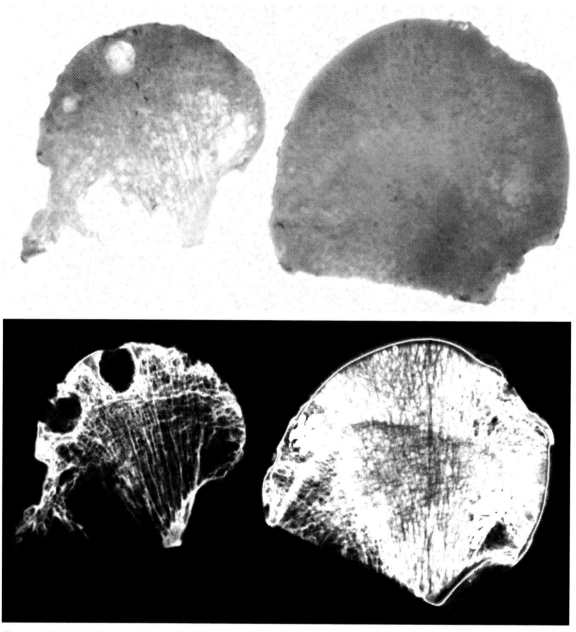

Figure 161a & b. Cross-section and radiograph of femoral head illustrating common indicators of osteoarthritis (left;) normal comparative specimen on right.

repeated trauma, infection (**Figure 160**), or avascular necrosis, among others (Hough and Sokoloff 1989). Resnick and Niwayama 1988; Schajowicz et al. 1979.

This facet (arrow) is most likely the result of contact of the tendon of rectus femoris on the neck of the femur (Kate 1963) and not compression of the antero-superior surface of the femoral neck by the

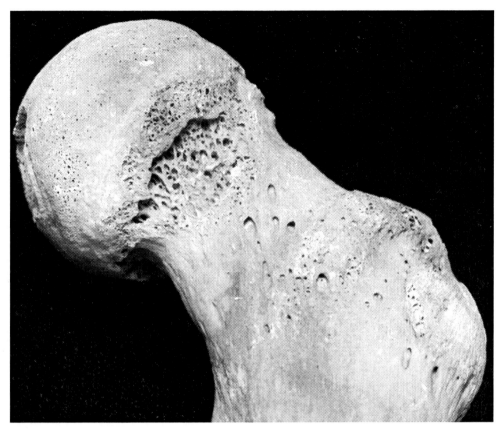

Figure 162. Anterior view of proximal femur showing a cervical fossa of Allen.

acetabulum, the so-called anterior acetabular imprint (Meyer 1924) (**Figure 162**). It is worthy of note that similar fossae have not been noted in habitually squatting nonhuman primates. Kate suggested that this fossa is the evolutionary result of the erect posture and is not due to squatting as others have stated (Molleson et al. 1998). Allen's (1882) original description of the femur neck stated that it "is marked in front near the articular surface by a faint depression, which is often cribiform in appearance and may receive the name cervical fossa." Findings by subsequent researchers have since made it difficult to distinguish one facet from another with the addition of such terms as the imprint of Bertaux (1891), Testut (1911), Poirier's facet (1911), and cribra femoris (Molleson et al. 1998). There are a variety of depressions, raised plaque-like areas and extensions of the articular surface (Poirier's facet) (**Figure 163**), and cribiform (porous/trabecular "sieve-like" [Allen's fossa]) lesions that may be present at various locations on the femur neck, possibly reflecting the bursa for crossing of the psoas tendon across the femoral neck. Allen's fossa can be distinguished from Porier's facet in that the former appears as if a "window" of the cortex has been cut out of the femur, revealing the underlying trabeculae. Look for other skeletal "excavations" at the attachment sites of the latissimus dorsi, teres and pectoralis in the proximal humerus, medial head of the gastrocnemius in the distal femur, and the costoclavicular ligament in the sternal end of the clavicle. These excavations are often found together in adolescents and young adults engaged in strenuous activities (such as soldiers) and may result from the same biomechanical or growth stress. Common finding, especially in adolescent and young adult males. Finnegan and Faust 1974; Kostick 1963; Meyer 1924; Saunders 1978.

An area of smooth or slightly raised bone, often about the size and shape of a thumbprint, along the anterior margin of the femur in the area of the bursa for the psoas tendon. This facet often appears as an extension of the articular surface of the femur head (**Figure 163**).

A common occurrence in the elderly ("broken hip"), especially postmenopausal and senescent females (**Figure 164**). Fractures of the femoral neck can result from acute trauma such as a fall or in the aged, with acute osteoporosis and act as simple as sitting down or lifting one-

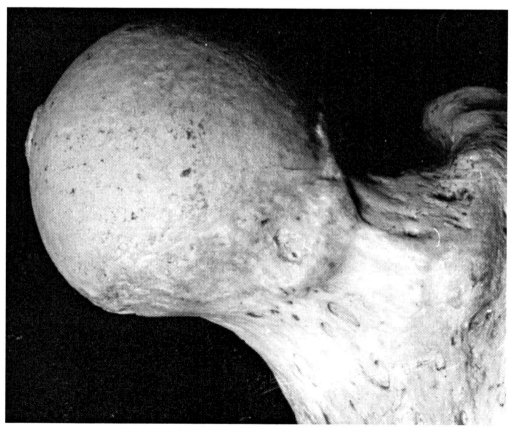

Figure 163. Kostick's (1963) or Porier's facet (1911).

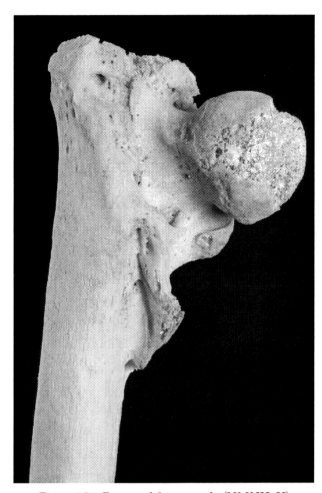

Figure 164. Fractured femur neck. (NMNH–H)

self from a chair. Fractured femur necks often lead to avascular necrosis (bone death). Death of the individual can be the ultimate result of a femoral neck fracture by disability and complications by immobility (e.g., pneumonia). In contemporary populations, 10–20 percent of elderly people who sustain "hip" fractures die with 6 months of injury (Genant 1989). Cummings et al. 1985; Gardner 1965.

Note the exuberant bone formation (often referred to as an exostosis), which would have extended into the muscles of the upper thigh. Heterotopic bone such as this (i.e., myositis ossificans) results from trauma to the muscle and subsequent ossification (**Figures 165–167**). Resnick and Niwayama 1988; Schajowicz 1981. (See Figure 180 for comparison.)

Marginal osteophytes. Raised, sharp lipping indicative of OA that may be the result of acute trauma (e.g., knee injury) or associated with old age (**Figure 168**).

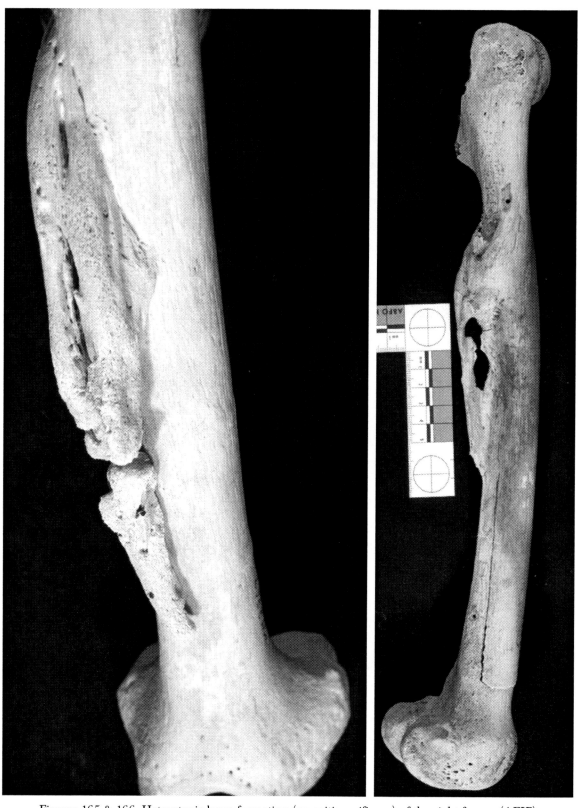

Figures 165 & 166. Heterotopic bone formation (myositis ossificans) of the right femur. (AFIP)

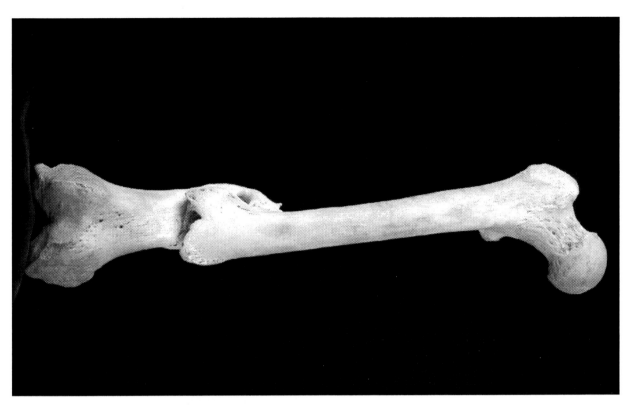

Figure 167. Healed fracture of the femoral diaphysis. Note angulation of the femoral shaft as a result of misalignment of the broken bones. (NMNH–H 321053)

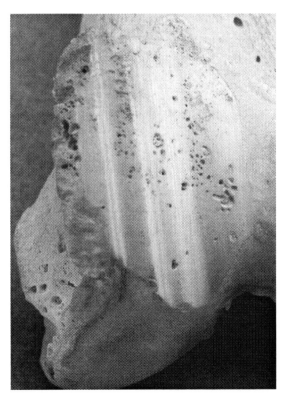

Figure 168. Severe OA/DJD of the distal femur showing marginal osteophytes and pitting of the articular surface with grooving and eburnation of the patellar surface. (NMNH)

Articular surface porosity. Small to large pits accompanying old age.

Articular surface osteophyte. Raised areas of bone on the articular surface.

Eburnation, resulting from the loss of cartilage and polishing ("shiny") of the articular surface from "bone on bone" wear. Eburnation of the femur is often accompanied by similar DJD of the patella and, possibly, the proximal tibia.

Osteophytes and macroporosity are common findings in the elderly. Eburnation is less common, and more often associated with more acute trauma to a joint where necrosis has occurred. Aufderheide and Rodriguez-Martin 1998:94–96; Rogers and Waldron 1995:38; Steinbock 1976:282–284.

These areas of rapidly growing and remodeling bone (**Figures 169–171**) are often misinterpreted as "diseased." See discussion concerning

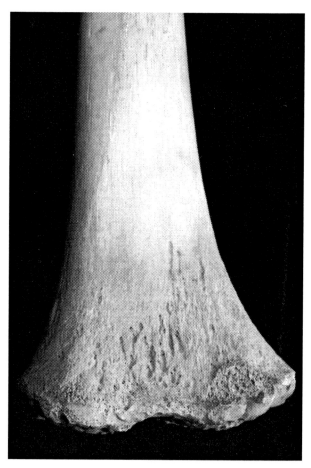

Figure 169. Normal growth areas (pitting, striations and discoloration) in the surface (dorsal) of a child's distal femur. (NMNH–H)

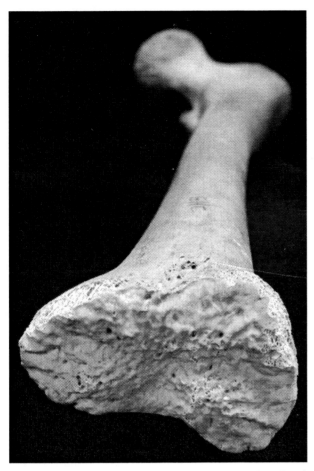

Figure 170. Normal metaphyseal plate in the distal femur of a child. (NMNH–H)

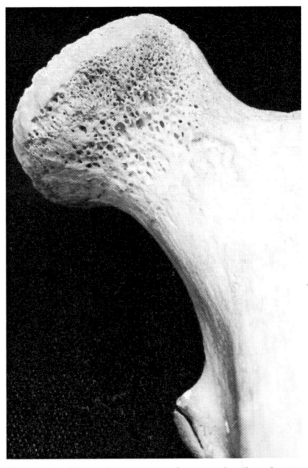

Figure 171. Normal porosity and exposed trabeculae in the femoral neck of a child (the femoral head was still unfused to the neck). (NMNH–H)

the "cut back" zone in normal bone growth of children presented by Ortner (2003:15–16).

Harris lines or growth arrest lines (Figures 172 & 173) appear as thickened lines in dry bone and opaque/sclerotic lines (radiographically) extending across the diaphysis and/or metaphysis of long bones (and teeth as well) as result of additional osseous deposition due to the failure of continuous growth at the metaphyseal cartilage. Note the wavy horizontal lines visible in the metaphysis in **Figure 172** and multiple lines in the distal ends of the humerus and femur and proximal end of the humerus in **Figure 173**. Controversy exists over the etiology of Harris lines as individuals never experiencing fever or other serious childhood illnesses or physiological disturbances have been shown to have these lines (Zimmerman and Kelley 1982). Gindhart (1969) reports that only 25 percent of severe diseases in immature subjects result in the formation of Harris lines. Harris lines, which may be vis-

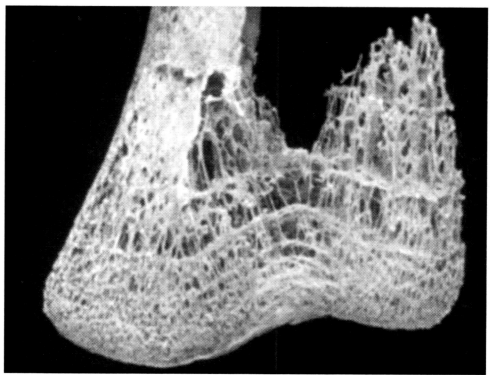

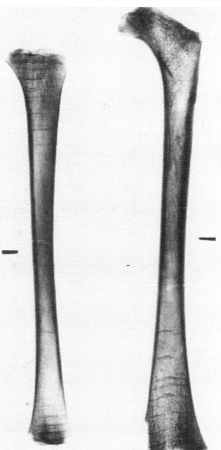

Figures 172 & 173. Harris lines (growth arrest lines, transverse lines, Park lines, Park-Harris lines, stress lines) indicative of multiple bouts of arrested growth (stress) during the period of growth as a subadult. (Areas of missing bone in the present examples are due to post-mortem breakage.)

ible in any of the long bones, may resorb over time leaving no evidence of these transverse lines of arrested growth. Aufderheide and Rodriguez-Martin 1998:422–424; Byers (1991); Goodman and Clark 1981; Grolleau-Raoux et al. 1997; Harris 1926, 1931, 1933; Lee and Mehlman 2003; Nowak and Piontek 2002a and b; Park 1964; Rosa 1994; Steinbock 1976:43–55.

A crater-like defect that can measure more than 2 cm in diameter, located above the medial condyle at the attachment of the medial head of the gastrocnemius muscle (**Figure 174**). The floor and margin of these defects often reveal sharp bony spicules and exposed trabeculae as though a plug of cortical bone has been forcibly avulsed. While the pathogenesis of this usually bilateral defect is unclear, most researchers believe it is due to the repeated pulling stresses (traction) or acute trauma of the gastrocnemius muscle, hyperemia (increased blood velocity), and localized bone resorption brought on by the traumatic event (Donnelly et al. 1999; Posch and Puckett 1998; Resnick 2002; p. 4563). A common finding of limited existence in children and

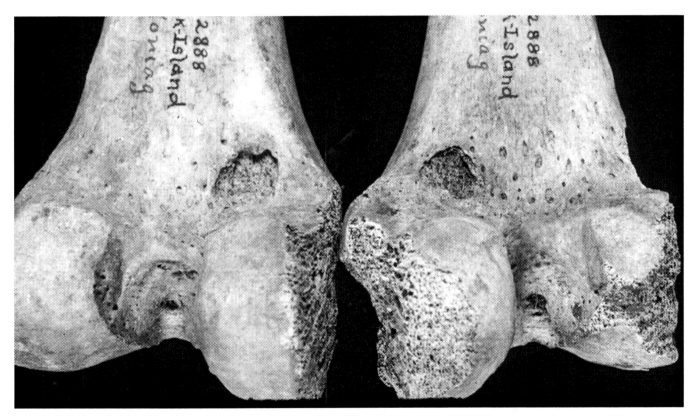

Figure 174. Distal femoral cortical excavation (DFCE) (Resnick and Greenway 1982), (subperiosteal cortical defect, cortical defect, cortical desmoid), ("tendon lesion," Hrdlicka 1914), avulsive cortical irregularity (Posch and Puckett 1998), fibrous metaphyseal defects (Ritschl et al. 1988, 1989). (NMNH 372888)

Figure 175a, b, c, & d. Left femur and patella (from below) showing bony changes due to prolonged flexion contractures of the lower leg (often due to poliomyelitis). (NMNH–T 781)

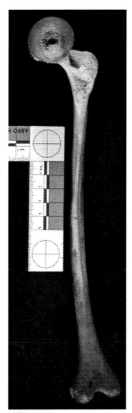

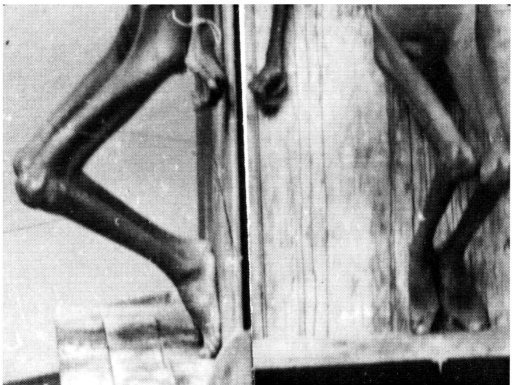

Figure 175 c & d. Note wasting of the femoral shaft (c) and "knobby" knees (d).

adolescents that is likely related to rapid bone remodeling during youth. Hrdlicka (1914) noted its presence in the skeletons of ancient Peruvian adolescents but not in children. The present authors have seen this feature in many subadults and adolescents, especially ancient Egyptians, Native Americans and Civil War and War of 1812 soldiers.

Narrow and seemingly elevated lateral condyle relative to the medial condyle (**Figure 175**). The lateral condyle serves to guide and hold the quadriceps tendon in place.

Deep trochlear fossa (patellar surface) due to contact pressure erosion of the quadriceps tendon. The average depth of most adult trochleae is 5.2 mm but some may be as deep as 10 mm (Casscells 1979). Trochleae deeper than 10 mm may represent a pathological condition in humans reflecting long term flexion contractures of the legs (knees drawn up towards the chest in a fetal position) such as being bedridden due to disease or injury (e.g., paralysis; Mann et al. 1991) or bipedalism, as quadrupeds have deep fossae while those in bipeds are shallow. Condition may be unilateral or bilateral (Conner 1970). Abraham et al. 1977; Micheli et al. 1986.

Note the following bony changes:

Wide intercondylar fossa.

Large, raised, and widely spaced lateral articular facet.

Small, raised, and widely spaced medial articular facet. If changes of this type are noted in either the femur or patella, it is almost certain that the individual endured many years with the lower legs in flexion (e.g., legs drawn toward the chest), possibly due to paralysis (associated with the condition noted in B above). Increased contact pressure of the quadriceps tendon against the distal femur results in erosion of the trochlear surfaces. A number of conditions including poliomyelitis, spinal cord trauma, and spina bifida, can cause flexion contractures of the legs. The finding of deep trochlear fossae or widely spaced patellar facets suggests cultural implications of caring for the individual who might have been unable to properly care for him/herself. Rare finding in most archaeological samples and uncommon finding in contemporary populations.

Neuropathy may affect any joint but shows a predilection for the large weight-bearing joints such as the knee (**Figure 176**). Although the specific etiology is debated, it is likely that Charcot's joints (severe osteoarthritis) develop after sensory loss (underlying neurological defect) to a joint and continued joint stress (Hellman 1989). This example depicts a hypertrophic Charcot joint of the knee involving the femur, tibia, and patella. Note the massive exuberant new bone and misalignment of the femur and tibia (the midline axis has changed).

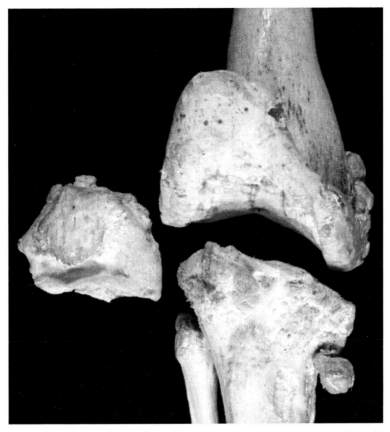

Figure 176. Neuropathic joint (Charcot's joint, anterior view of knee). (AFIP)

Conditions predisposing the development of a Charcot joint include alcoholism, diabetes mellitus, leprosy, congenital insensitivity to pain, and syringomyelia (Sella and Barrette 1999). Rare finding. Aufderheide and Rodriguez-Martin 1998:107–108; Brower and Allman 1981; Bryceson and Pfaltzgraff 1990; Eichenholtz 1966; Jaffe 1975; MacAusland and Mayo 1965; Ortner and Putschar 1985.

Rickets (**Figure 177**) is caused by the lack of calcium and phosphorous in the diet, or the inability to metabolize or absorb these minerals by intestinal maladies such as chronic diarrhea, metabolic insufficiency, renal failure, etc. See further discussion in Auferheide and Rodriguez-Martin 1998:305–310. Ortner 2003:393–404; Zimmerman and Kelley 1982:60.

This fossa, located along the postero-lateral surface of the femur and inferior to the lesser trochanter, appears as an elongated groove or shallow depression running parallel to the long axis of the femur (**Figure 178**).

Cloacae or draining sinuses (for pus) are frequent findings in chron-

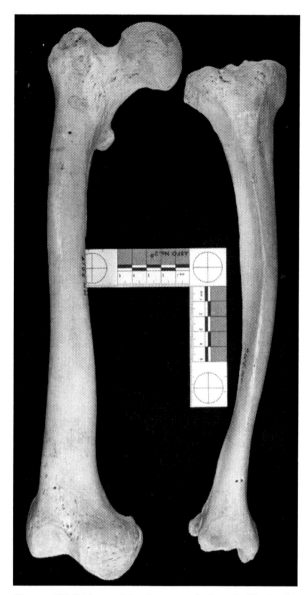

Figure 177. Rickets of the femur and tibia. Aufderheide and Rodriguez-Martin 1998:305–310; Mankin 1974a&b, 1990; Ortner 2003:393–404; Steinbock 1976:262–273; Zimmerman and Kelley 1982:58–60. (AFIP MM3913)

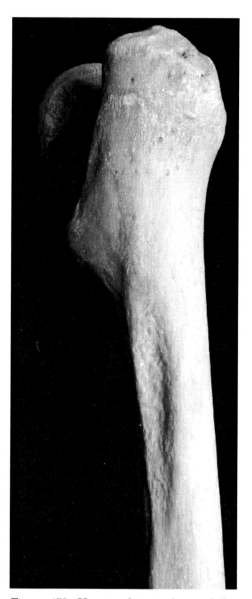

Figure 178. Hypotrochateric fossa of the femur (normal variant more commonly found in American Indians).

ic suppurative osteomyelitis (**Figures 179 & 189**). The light island of bone visible within the cloaca (in **Figure 189**) is a sequestrum, that is, and isolated dead bone fragment that became walled off by exudate, granulation tissue, or scar. The outer envelope of irregular bone is an involucrum (i.e., sheath of newly formed bone). See Ortner 2003:179–206 for numerous illustrations of osteomyelitic effects to bone (Figs. 9–1 through 9–32) and Steinbock (1976:60–85). As stated

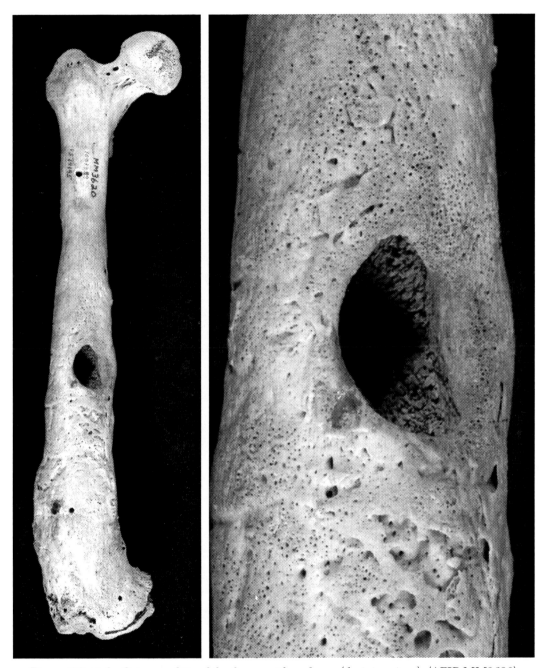

Figure 179a & b. Osteomyelitis of the femur with a cloaca (draining sinus). (AFIP MM3620)

by Carney and Wilson (1975), "The special pathologic tissue reactions which are characteristic of suppurative osteomyelitis are the lysis of bone, the formation of new bone, and the presence of dead bone." If the fragment(s) is small enough it will gradually be resorbed/absorbed). In dry bone the sequestrum may be freely mobile and rattles when shaken. Uncommon to rare finding in archaeological specimens.

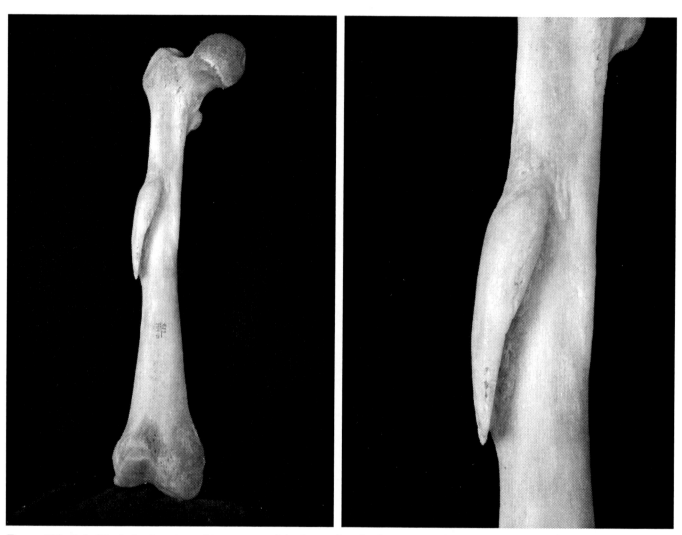

Figure 180a & b. Healed subperiosteal hematoma of the femur (similar lesions may be found in any bone). Day 1960; Kullman and Wouters 1972; Steinbock 1976. (NMNH–H)

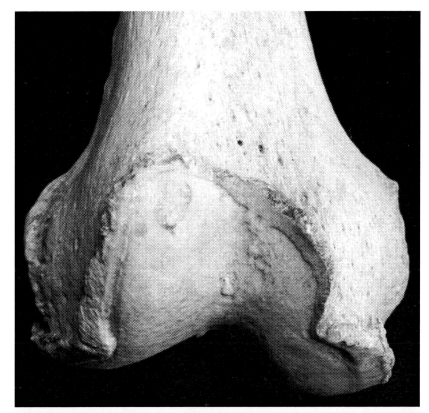

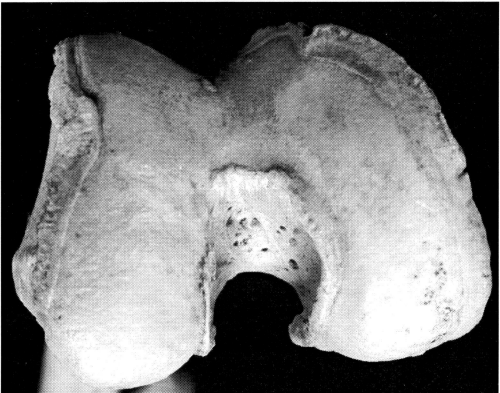

Figure 181a & b. Marginal osteophytes of the distal femur.

Figure 182. Moderate osteoarthritis of the femoral head and fovea capitis. The degree of severity in this example is commonly found in modern groups, but less so in ancient groups.

May be common in hospital and battle-related skeletal samples where surgical intervention prolonged and often exacerbated infection (e.g., Civil War soldiers). Aufderheide and Rodriguez-Martin 1998:174–177 (especially a complete sclerotic involucrum of a tibia at the Mutter Museum) (Fig. 7.45).

TIBIA

Periostosis results from inflammation stimulus from infectious activity or as the result of direct trauma such as a blow to the bone where the haematoma lifts the periosteum from the bone, stimulating bone deposition (**Figures 183 & 184**). One possible result of the latter is an ossified periosteal hematoma (Day 1960) (see **Fig. 180a & b**). Other causes for periostosis can be blood born infection (syphilis, tuberculosis), venous insufficiency as in varicose veins (Daniels and Nashel 1983; Gensburg et al. 1988), scurvy, or any other host of factors (Aufderheide and Rodriguez 1998:310–311). Common usage of the term *periostitis* generally refers to bone formed only on the outer cortex while osteitis refers to changes within the cortex, and osteomyelitis to changes that affect both the marrow and bone (Caraveo et al. 1977). Many times it is difficult, if not impossible, to ascertain the cause of the periosteal remodeling. Periostosis, in the active stage, is easily recognized by its color, texture, and raised appearance. Active periostosis (A) is often discolored from the underlying bone, pitted due to the increased vascularity, striated, and has defined, raised margins (B). Active periostosis has the look of small, loosely attached sheets of tree bark or thin layers of sponge. Levine et al. 2003; Ortner 2003:206–215.

In healed cases, the margins of the defect are less distinct, is less porous/pitted than in active cases, and "blend" into the surrounding bone. The rough striated or spongy morphology has begun to be remodeled, making a more smoothed appearance. It should be remembered that the periosteum in fetuses, newborns, and children is only loosely adhering to the cortex and can be easily lifted resulting in massive periosteal remodeling (e.g., congenital scurvy or syphilis; Toohey 1985). Ortner et al. 2001.

The tibial spines (**Figure 185**), although not ligamentous insertions but articular surfaces, will be slightly elongated and sharp to the touch. Common finding in the elderly.

Raised islands of bone on the articular surface (small arrow) or along the articular margin are bony responses to the destruction and degeneration of the joint capsule and cartilage covering the joint (large arrows). Marginal osteophytes appear as irregular, raised bony margins resulting from enchondral ossification that obliterate the articular margin (**Figure 186**). This is unlike syndesmophytes which are inflammatory spurs, and enthesophytes which are traction spurs at the insertion site of tendons and ligaments. This is a common finding associated with increasing age. Rogers et al. 1997.

Rounded, smooth or irregular bony projections, usually extending perpendicular to the long axis of the diaphysis. These growths may

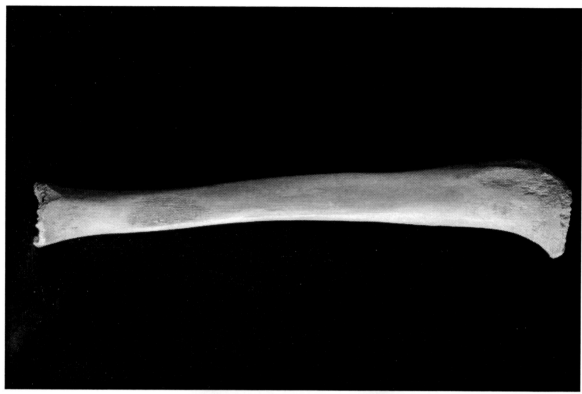

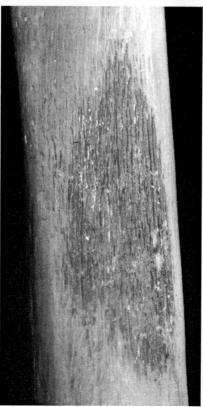

Figure 183a & b. Active periostosis (lateral view of right tibia). (NMNH–H)

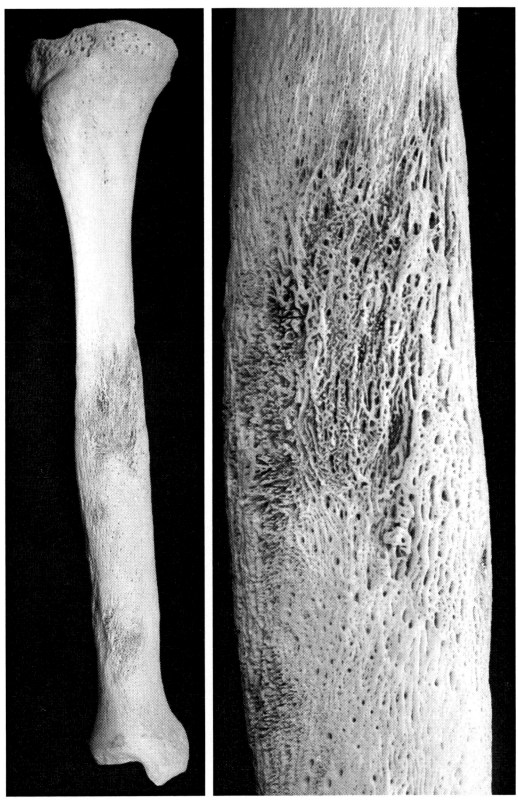

Figure 184a & b. Healed periostosis of the tibial diaphysis (mild forms are common in many skeletal groups and along any long bone). (NMNH–H)

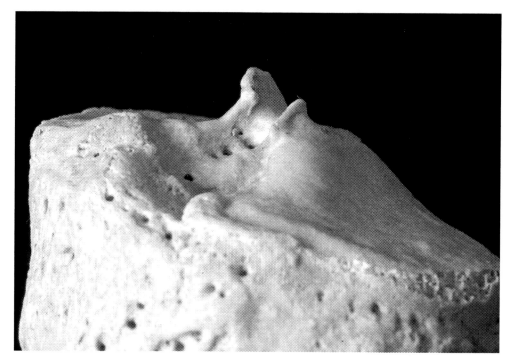

Figure 185. Spiking/peaking of the tibial spines–early indicator of osteoarthritis (Alexander 1990). (NMNH)

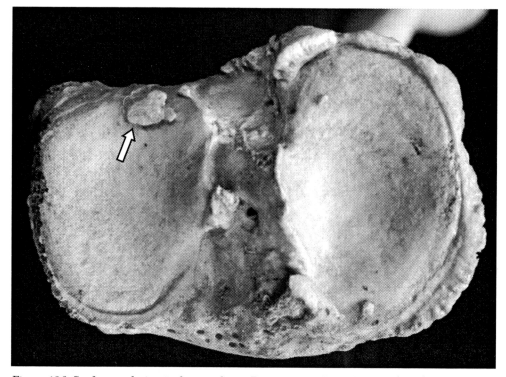

Figure 186. Surface and marginal osteophytes Degenerative Joint Disease (DJD) on the proximal tibia.

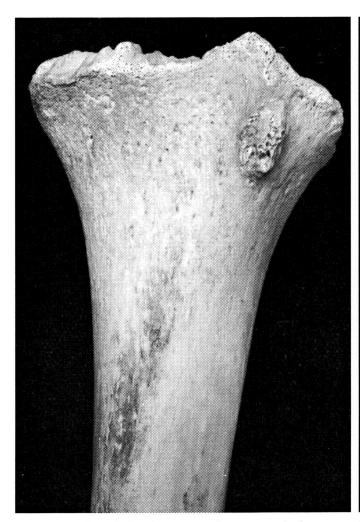

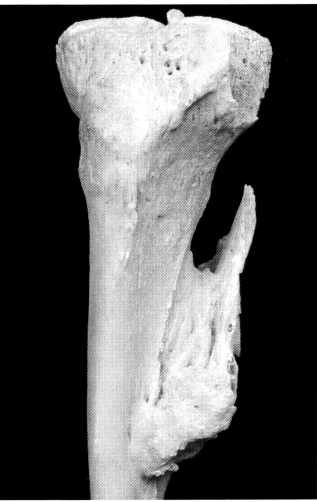

Figure 187. Normal development and presence of what some may mistake as a bone spur or osteochondroma (circle) in the proximal tibia of a child. (NMNH–H)

Figure 188. Exostosis along the interosseous crest of the tibia. (NMNH–H) (Compare with Figure 180)

reflect trauma to muscle or tendinous attachment or may be formed by benign ossous tumor (see **Figure 187** for normal bone and **Figure 188** for abnormal exostosis). Frequent sites of involvement are the metaphyseal areas of the shoulder, knee, and ankle (Spjut et al. 1971). Uncommon finding in most archeological samples. For more concise identification as to the cause of the formation, seek the advice of a radiologist and/or perform histological assessment. Lange et al. 1984; Ortner 2003:509–510, Fig. 20–8; Steinbock 1976:319–336.

Many diseases (particularly infectious) as well as congenital factors or trauma (especially fracture) and subsequent remodeling may result in bony union of two or more bones (**Figures 190 & 212**). The bones fuse through the proliferation of callus, osteophytes, or exostoses.

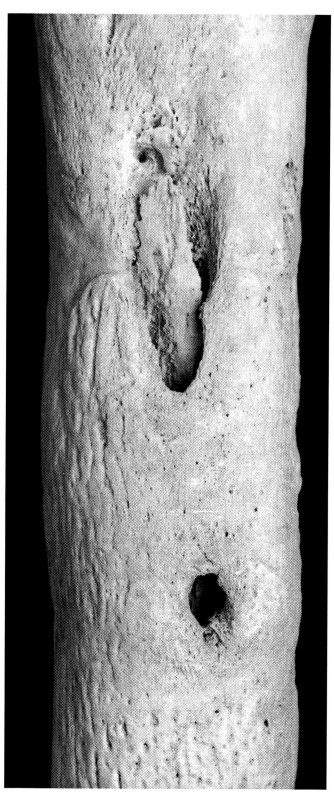

Figure 189. Osteomyelitis of a subadult tibia with sequestrum of bone and cloaca (see Ortner 2003:200–203, Fig. 9–29 for in-depth discussion of this specimen). (NMNH 378243)

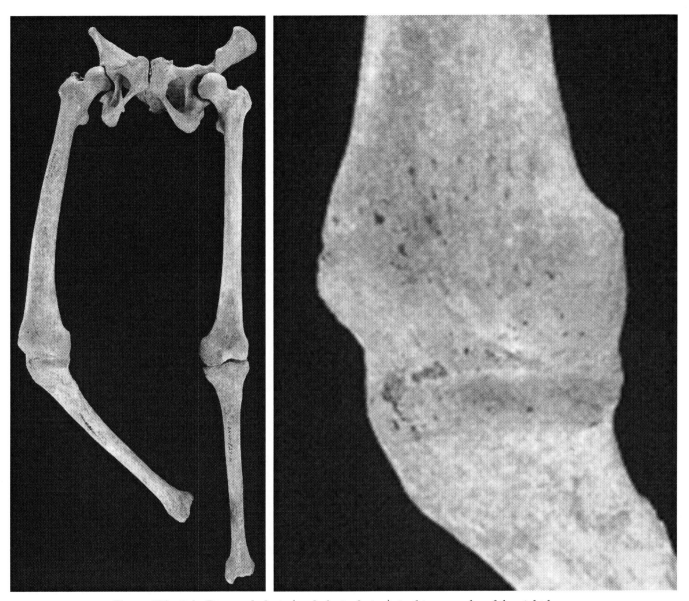

Figure 190a & b. Bony ankylosis (anchylosis, fusion), in this example, of the right knee.

Septic or pyogenic arthritis may fuse joints together, and illustration of involved knees are presented in Ortner (2003:255), Webb (1995:167) and in Zimmerman and Kelley (1982:95). Bony ankylosis of the joints of major long bones is uncommon to rare. When ankylosis of a joint is encountered, seek the advice of a specialist for determination of the cause. Ankylosis of the ribs to the vertebrae (AS) and bones of the hands and feet, especially the middle and distal phalanges of the feet is an uncommon to common finding. Bony ankylosis of joints is rarely the result of osteoarthritis (Lasater and Groer 2000).

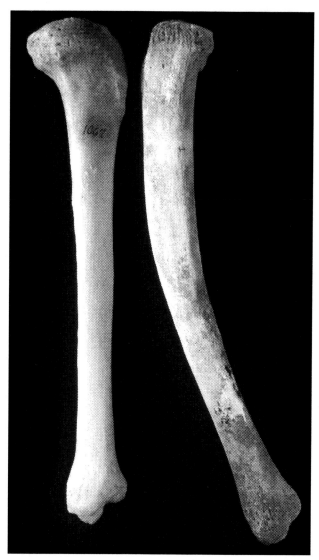

Figure 191. Sabre-shin tibia ("boomerang leg") compared to normal tibia.

Tibiae affected by acquired or congenital syphilis (**Figures 191 & 192**) or some of the other treponematoses (e.g., yaws in tropical areas) may exhibit periostosis (A) along the anterior and middle portion of the shaft, resulting in an anterior enlargement ("bowing" or sabre-shin) of the bone. Sabre-shin tibiae were noted in 9 of 100 (Caribbean) and 11 of 271 (native-born Americans in Boston) patients with late congenital syphilis (Fiumara and Lessell 1983 and 1970 respectively). Pavithran (1990) has identified sabre-shin features in leprosy. Andersen 1991; Aufderheide and Rodriguez-Martin 1998:159–160; Goff 1967; Hackett 1975, 1976; Krogman 1940; Mays 1998; Mays et al. 2003; Ortner 2003:294–296; Rasool and Govender 1989; Steinbock 1976:98–158.

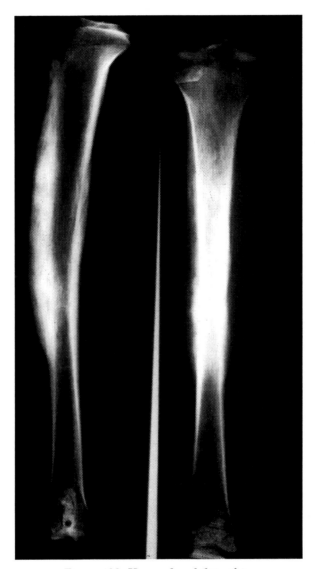

Figure 192. X-ray of syphilitic tibia.

Note the presence of new bone (radio-opaque (whitish) area) along the proximal diaphysis of the tibia giving it the characteristic "sabre-shin" or "boomerang" shape.

A squatting facet (**Figures 193 & 194**) is scored as present whenever there is a break in the continuity of the anterior margin of the distal tibia, which should be a straight line. The anterior surface of the tibia abuts the dorsal portion of the talar neck when the knee and foot are are hyperdorsiflexed and are usually attributed to the squatting position (Boulle 2001; Jeyasingh et al. 1979; Oygucu et al. 1998; Pandey and Singh 1990). Massada (1991) found a high incidence of squatting facets in soccer players. There may be one or two facets

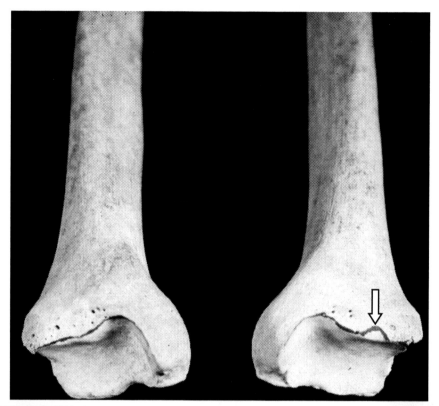

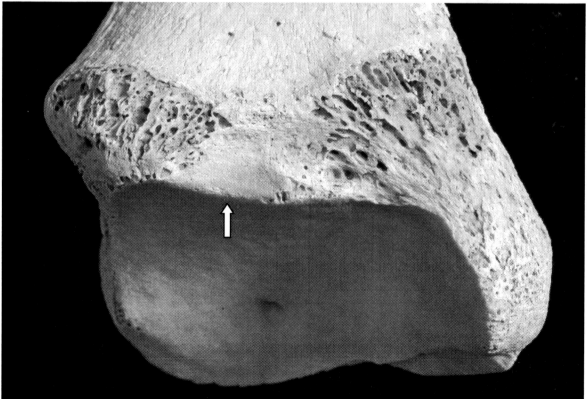

Figures 193 & 194. Squatting facets of the distal tibia. (Fig. 194–NMNH 293624)

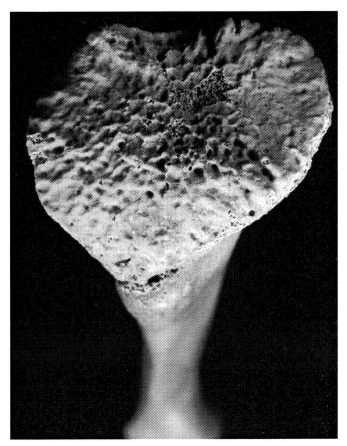

Figure 195. Proximal tibia, showing normal metaphyseal growth plate. Note the irregular/billowy surface. (NMNH–H)

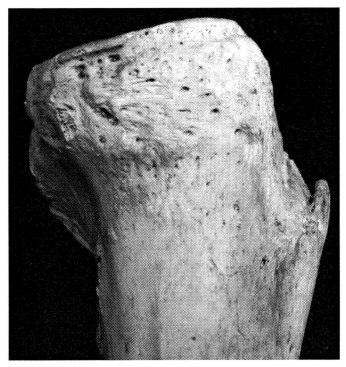

Figure 196. Enthesophytes of the tibial tuberosity. (Mild cases are commonly encountered in most skeletal groups studied.) (NMNH–H)

(medial and lateral) and are common findings in most if not all populations. Squatting facets may also be present on the talus, calcaneus, distal metatarsals, and proximal tibia. Barnett 1954; Kate and Robert 1965; Lima et al. 1928; Ubelaker 1999:103.

Bony spurs are formed by the ossification of the patellar tendon, most often from trauma (Resnick 2002:3589). This outgrowth (**Figures 195 & 196**) may also be associated with Osgood-Schlatter's disease but diagnosis can be tentative (D'Ambrosia and MacDonald 1975). Aufderheide and Rodriguez-Martin 1998:85; Ortner 2003:353–354 (see Fig. 13–9 for illustration of Osgood-Schlatter's with radiograph).

Figure 197. Pott's fracture (Dupuytren fracture) of the distal fibula with exostosis of the interosseous ligament.

FIBULA

Pott's fracture, first described by Percival Pott in 1769, is loosely used to refer to fractures and dislocations of the distal tibia and/or fibula occurring 2 to 3 inches above the ankle joint (Apley and Solomon 1988). The type of fracture (**Figure 197**) is dependent on the type and direction of force (e.g., lateral impact) exerted to the lower leg. For example, eversion and external rotation of the lower leg can result in fracture of the lateral or medial malleolus, with rupture of the interosseous ligaments. This tramatic rotation of the ankle (especially by the talus) will separate the fibula from the tibia, tearing the interosseous ligaments, resulting in hemorrhage which can lead to subsequent ossification of these ligaments. In most groups, common finding in adults and rare in children. See Sammarco and Cooper (1998) for fracture classification types using the Lauge-Hansen or Weber system. Adelaar 1999; Brown et al. 2003; Hamilton 1984; Jahss 1991; Ortner 2003:158; Resnick and Niwayama 1988; Sisk 1987; Wilson 1975.

PATELLA

A condition in which the patella appears as to have a notch taken out of the superolateral corner (most frequently observed location) at the insertion of the vastus lateralis muscle (**Figure 198**). Halpern and

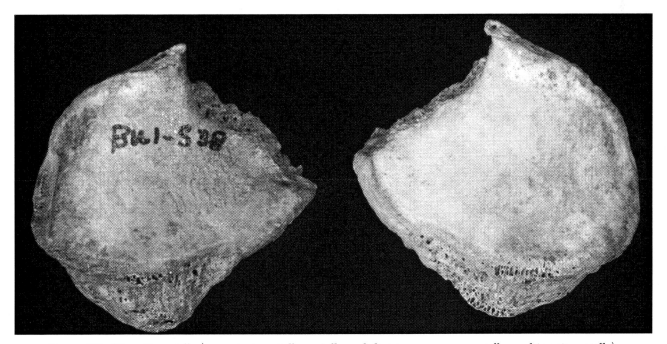

Figure 198. Bipartite patella (emarginate patella, patellar subdivision, accessory patella, multipartite patella).

Oakley (1978) reported a possibly unique case of bilateral medial bipartite patellae. The notch is usually located at the site for attachment of a smaller accessory ossicle, the so-called "patellula" in the older literature (Oetteking 1922).

Although the etiology of this condition is unknown, it is commonly believed to represent an accessory center of ossification and not the result of direct or indirect trauma to the patella (Green 1975; Halpern and Hewitt 1978; Miles 1975; Ogden et al. 1982). Bipartite patellae occur in approximately 2 percent of the population (George 1935), are more common in males by 9 to 1 (Ogden et al. 1982), and are unilateral in 57 percent of the cases (Weaver 1977). Pain may or may not have been associated with this condition. (See Yochum and Rowe [1996] for a good x-ray example of this feature.) Stanitski 1993.

A review of the literature revealed that some researchers make no distinction between a bipartite patella and vastus notch (E) (see **Fig. 204**). Some early anatomists (Kempson 1902; Todd and McCally 1921) viewed defects of the superolateral border as manifestations of a single anomalous condition (emargination) with a vastus notch representing the mildest form, and a bipartite patella the most severe form. Most researchers, however, now distinguish between a vastus notch and bipartite patella. A bipartite patella has a porous, roughened central area surrounded by smooth-bordered cortical bone for attachment of an accessory (bipartite) bone. A vastus notch, on the other hand, is a smooth surfaced, depressed or flattened area with no accompanying porosity. Bipartite patella is a common clinical, but uncommon archaeological finding. Vastus notches are very common. Adams and Leonard 1925; Callahan 1948; Giles 1928; Jones and Hedrick 1942; Miles 1975; Pecina and Bojanic 2004; Resnick 2002; Saunders 1978; Wright 1904; Zimmerman and Kelley 1982 (p. 36, Fig. 19).

The posterior surface of the patella may exhibit a raised rim at any point along its margin (**Figures 199–201**). Other indicators of osteoarthritis (OA) are surface osteophytes or surface porosity and pitting, usually present in the medial facet. OA of the patella is a common finding in the elderly, but is usually first evident in the late thirties (Todd and McCally 1921). OA in young individuals may reflect faulty/soft cartilage known as chondromalacia or shearing stress and cartilage/osseous degeneration due to acute trauma particularly with the knee flexed (Edwards and Bentley 1977). Normal patellae exhibit smooth, sometimes undulating facets for articulation with the distal femur. Large, widely separated and raised oval facets may indicate paralysis and flexion contractures (bent knees) of the legs (**Figure 175**). This feature is uncommon.

Osteoarthritis (OA) can be distinguished from osteochondritic pits

Figure 199. Marginal osteoarthritic "lipping" of the patellar facets. (NMNH–H)

and other normal variants in that pitting reflecting arthritis is usually characterized by irregular, jagged margins.

Note the vertical orientation of grooves (**Figure 201**) from contact with the distal femur (in this case, the femur exhibited similar grooves). Also note the presence of marginal osteophytes (a mild to moderate degree of severity inconsistent to the severity of the eburnation) encircling the patella (**Figure 202**).

Elongation and projection of the inferior border of the patella resulting from pulling stresses at the inferior ossification center of the immature patella by the patellar tendon (Smillie 1962). Uncommon finding.

Exuberant bone seemingly attached to and flowing across the anterior surface of the patella (**Figures 202 & 203**). Small bony projections

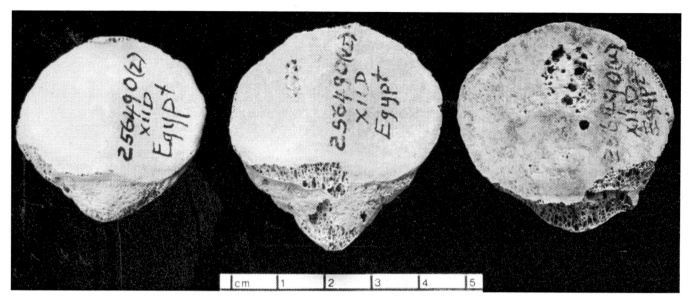

Figure 200. "Pitting" type of osteoarthritis on the patellar facets (normal = left, middle = mild/moderate, right = moderate /severe). (NMNH 256490)

Figure 201. Eburnated articular surface due to destruction and loss of articular cartilage at the knee. (NMNH-H)

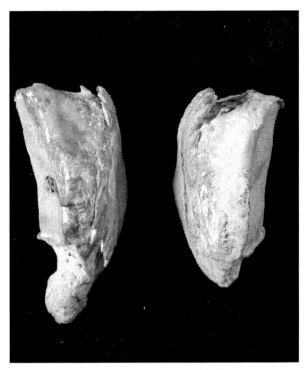

Figure 202. Elongated inferior pole (left) and entheso-phytes at the insertion of the quadriceps tendon.

Figure 203. Ossified quadriceps tendon (enthesophytes).

(enthesophytes) on the superoanterior face of the patella are common findings in most populations. However, development as shown in **Figures 202 & 203** is more likely to be seen in the elderly, typically beyond 60 years of age and likely reflects DISH. Usually bilateral and symmetrical in appearance. If there is spiking of only the superoanterior surface of the patella the condition reflects enthesopathy, not ossification of the quadriceps tendon or osteoarthritis. Dieppe et al. 1986; Resnick and Niwayama 1988.

This non-metric variant appears as notch or depression in either the superolateral (commonly) or superomedial corner of the patella (**Figure 204**). The vastus notch can be distinguished from the bipartite patella in that the surface of the vastus notch is smooth and nonporous. The shape of the notch varies from nearly flat to concave. Common finding. Finnegan and Faust 1974.

Edwards and Bently (1977) noted Osteochrondritis dissecans (OD) in the patellae of three males (one bilateral) and two females 13 to 17

Figure 204. Vastus notch (emargination).

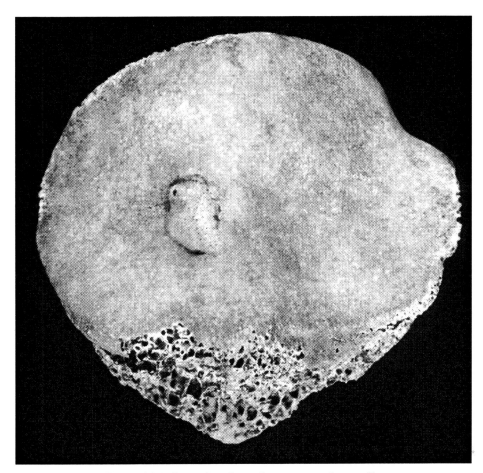

Figure 205. Osteochondritis dissecans on the patella.

years of age (**Figures 205, 206 & 218**). Desai and colleagues (1987) studied 13 cases of OD in nine male and 2 female patients between the ages of 10 and 26 years. Their study confirmed that most cases developed in the second and third decades of life, were found predominately in males, and that all of the defects were located in the lateral facet and they found none in the medial facet. They concluded that OD of the patella results from minor repeated trauma to the articular patellar surface as none of the patients reported any history of major injury to the knee. Outerbridge 1964.

The term osteochondritis is used to cover a large group of conditions (over 80 different types) that predominately affects overweight children between the ages of 5 and 11 years (Griffiths 1981). Uncommon finding in archaeological samples. For an interesting discussion of the etiology of OD see Barrie (1987). Aufderheide and Rodriguez-Martin 1998:81–83. Ortner 2003:351–353. See illustrations for OD in other bone articulations in this volume.

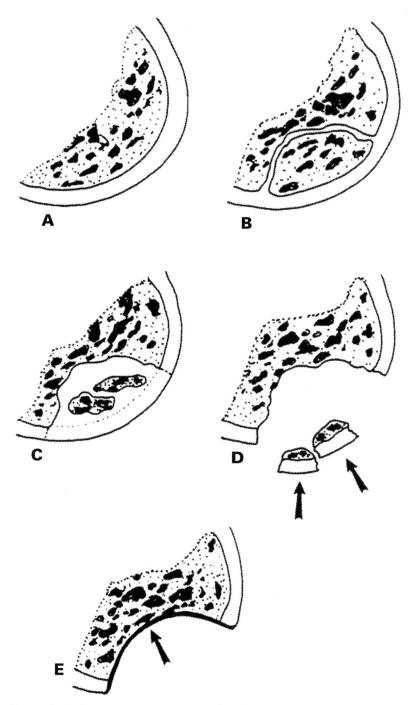

Figure 206. Osteochondritis dissecans (OD): drawings of the stages and outcome of the condition (reprinted with permission from Harry J. Griffiths, *Basic Bone Radiology,* Appleton-Century-Crofts, Norwalk, CT, 1981, and modified). (**Figures 156, 200, 206 & 218**)

Note the following progressive bony changes:
Normal joint.
Infarcted area with separation of a large bone and cartilage fragment.
Resorption of the contents and showing future line of weakness in cartilage.
Formation of loose bodies ("joint mice"—arrows).
With time the defect (osteochondritic pit) heals and appears as a smooth-walled crater (arrow).

FOOT

Os trigonum is a separate small bone (accessory ossicle) of the posterior tubercle of the talus (**Figure 207**). Some researchers identify an os trigonum as such even if the bone fragment is not separate, but exhibits a fissure in the undersurface of the posterior talar facet (Steida's or trigonal process; Sarrafian 1993). In a study of 1000 radi-

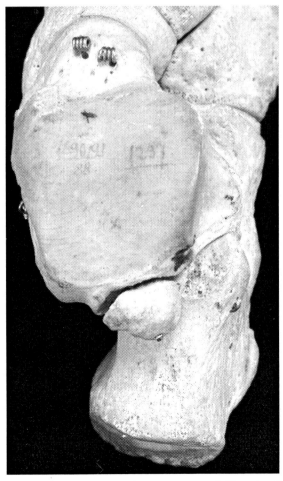

Figure 207. Os trigonum (non-metric trait) (Dwight 1906).

ographs by Burman and Lapidus (1931), the os trigonum was found in nearly 50 percent of the feet. A fused os trigonum, however, may fracture and be accompanied by pain and discomfort–especially when the foot is plantar flexed in such activities as ballet, American football and soccer–and become a free bone (Ihle and Cochran 1982; Lombardi et al. 1999; Marotta and Micheli 1992; Oloff et al. 2001). The examiner must decide which criteria to use in scoring this trait as present or absent. The present authors prefer to score this trait as present only when a separate bone is present (if still attached it is a Steida's process). An os trigonum, even in the absence of a loose bone/fragment, may be detected by the presence of porous bone encircled with a bony margin on the distal portion of the talus. Brown et al. 1995; Karasick and Schweitzer 1996; Mann and Owsley 1990; McDougall 1955; Resnick 2002; Sarrafian 1993; Yochum and Rowe 1996.

Although the etiology of congenital clubfoot is unknown (Altchek

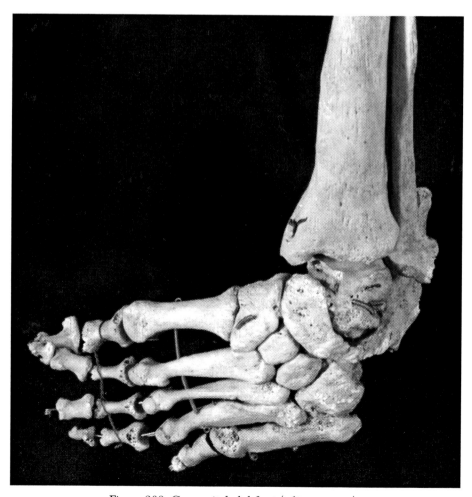

Figure 208. Congenital clubfoot (talipes or pes).

1978), the resulting foot deformity (**Figure 208**) is due to malformation of the talus and medial rotation of the calcaneus. There are a number of congenital deformities of the foot including talipes equinovarus in which the foot is rotated medially and the individual walks on the "outside" of the foot. Talipes equinovarus is the most common major congenital deformity occurring once in every 1000 contemporary births (Kromberg and Jenkins 1982; Shands 1951) and males are afflicted nearly twice as often as females (Salter 1970). Other clubfoot deformities include talipes equinus (heel off the ground and walking on the toes), talipes valgus (the heel and foot are turned outward), and talipes varus (the heel is turned inward and away from the midline (Thomas 1985). The incidence of congenital clubfoot in Asians is less than that of Whites (Morton 1970; Niswander et al. 1975; Wynne-Davies 1965). Mann and Owsley 1989; Owsley and Mann 1990; Ponseti 1994; Ponseti et al. 1981.

Figure 208 depicts the congenitally malformed left foot of an individual with severe talipes equinovarus. Note that the proximal ends of the metatarsals are overlapping and the distal ends fan out. When walking, the individual would have placed most of the body weight on the outer side of the fifth metatarsal (which is evidenced as thickened

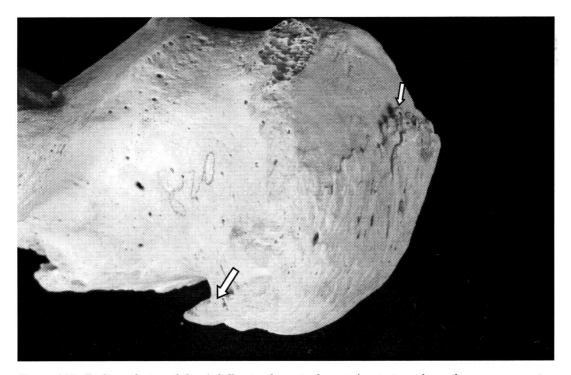

Figure 209. Enthesophytes of the Achilles tendon attachment (posterior calcaneal spur or posterior spur) (small arrow). Enthesophyte (inferior calcaneal spur, heel spur) (large arrow). (NMNH 256494h)

radiographically). Skin abrasions and subsequent infection often accompany untreated clubfoot (Laura Tosi pers. comm. 1989). In severe cases individuals may actually walk on the distal end (malleolus) of the tibiae.

Bone spurs at this site (**Figure 209**) are due to repeated or acute trauma (bleeding and inflammation) of the Achilles tendon. Common finding. There also may be erosion and bony projections at this site indicative of Reiter's disease, an uncommon to rare finding (consult a radiologist). Resnick et al. (1977) reported finding 16 percent of individuals over age 50 had plantar spurs (large arrow, **Fig. 209**), 11 percent had posterior spurs (small arrow) and 4 percent had both. Banadda et al. (1992) found that calcaneal heel spurs were common in the general population but showed increasing frequencies with age (10–16% of individuals over age 50). Bassiouni 1965; Dieppe et al. 1986; Hertzler 1931; Kumai and Benjamin 2002; Lehman 1999; Lu et al. 1996; Resnick and Niwayama 1988; Riepert et al. 1995; Thomas et al. 2001; Turlik 1990.

Bony projections resulting from repeated or acute trauma to the abductor hallucis and flexor digitorum brevis tendon attachments. Stresses at this site result in the lifting of the periosteum with concomitant inflammation and bone formation. Pain is usually only associated with inferior spurs during the inflammatory stage (i.e., when the trauma occurred, Michael R. Droulette pers. comm. 1989). The inclination of the calcaneus is such that inferior spurs are nonweightbearing and do not come in contact with the ground. Calcaneal spurs show no sex predilection, but the frequency of occurrence increases with age, especially over age 50. D'Ambrosia 1987; Hough and Sokoloff 1989; Resnick et al. 1977; Thomas et al. 2001. Common finding in the elderly.

An accessory bone (secondary center of ossification) that coalesces with the anterior calcaneal facet that, in the living, is separated from the calcaneus by a zone of cartilage. Normal variations of the talar articular facets of the calcaneus include two separate facets (middle and anterior) of approximately equal size and shape, or one large middle facet and one small anterior talar articular facet, or one large middle facet and one rudimentary or nonexistent anterior facet. Although the actual bone fragment of a calcaneus secundarius (**Figure 210**) is rarely recovered with archaeological remains, the notch at the anterior aspect of the calcaneous will be visible. The notch will be concave, roughened, and porous. The presence of porosity serves to distinguish a calcaneus secundarius from normal variation in the shape of the anterior calcaneal facet. Research by one of the authors (RWM) revealed that this trait is probably present in all populations with a fre-

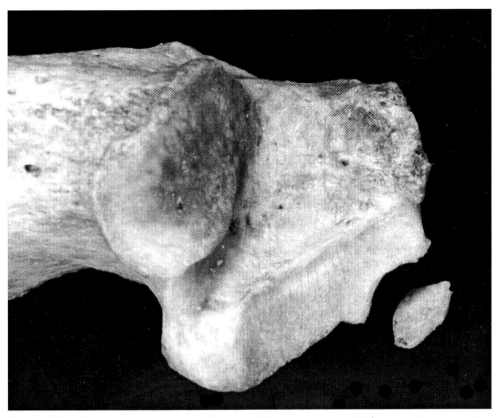

Figure 210. Calcaneus secundarius (Gruber 1871; Anderson 1988), secondary os calcis (Dwight 1906).

quency of between 1.4 and 6.0 percent (Mann 1989, 1990). Care must be exercised not to confuse this normal detached ossicle with a fracture and loose bone (Hodge 1999).

Osseous or cartilaginous coalition of the talus and calcaneus (**Figure 211**), usually of the middle facet of the sustentacumlum tali, is an inherited embryonic anomaly resulting from a failure of primitive mesenchyme to segment in the fetus (Kawashima and Uhthoff 1990; Richardson 2000). Talocalcaneal middle facet and calcaneonavicular coalitions are the most common of the tarsal coalitions (Percy and Mann 1988). Tarsal coalition has been likened to a "spot weld" of a joint and is found more often in males. This coalition is often associated with talar beaking and is found bilaterally in 50 percent of cases. There appears to be no population specificity and is asymptomatic in most adults, or accompanied by minor pain and stiffness. Beckly et al. 1975; Bhalaik et al. 2002; Bohne 2001; Solomon et al. 2003.

This condition (**Figure 213**) results from a combination of genetics, repetitive biomechanical stress, and an immune response to bacterial products leading to inflammatory changes at specific anatomical sites

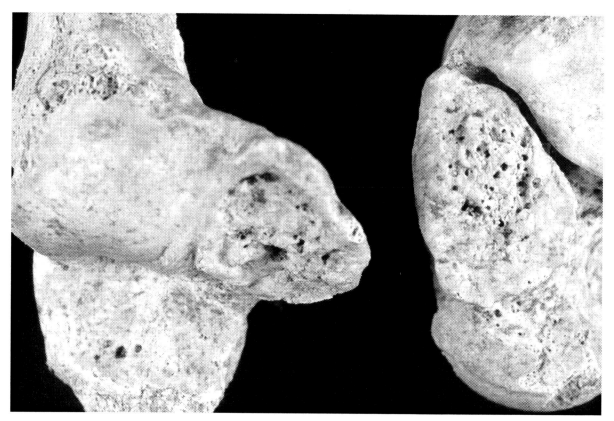

Figure 211. Talocalcaneal coalition.

and joints. The spondyloarthropathies include psoriatic arthritis and ankylosing spondylitis, among others. Bernhard 2002; Deesomchok and Tumrasvin 1993; Luong and Salonen 2000; Steinbock 1976:304–312.

Note the destruction (acro-osteolysis) of the metatarsals and phalanges. Many of the foot bones become "penciled," cupped and pegged (metatarsals and phalanges), deformed, and misaligned (**Figure 214**). The nutrient foramina, especially those of the phalanges may be enlarged. Frostbite and rheumatoid arthritis, however, can result in similar deformities and disintegration. The tibia and fibula may also become involved and exhibit bilateral symmetrical inflammatory changes including pitting and marked periosteal new bone formation (Manchester 1989). If this condition is encountered, seek the advice of a specialist. Andersen 1969, 1991; Anderson 1982; Aufderheide and Rodriguez-Martin 1998:141–154 (epidemiology and geographic distribution as well as lesions and their affects on the bone are discussed); Boldsen 2001; Burgener and Kormano 1991; Harverson and Warren 1979; Kulkarni and Mehta 1983; Møller-Christensen 1953, 1961; see Ortner 2003:263–271 for overview of

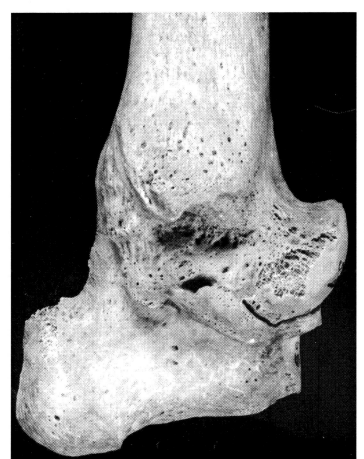

Figure 212. Ankylosis (fusion) of the tibia with the talus and calcaneus. (AFIP)

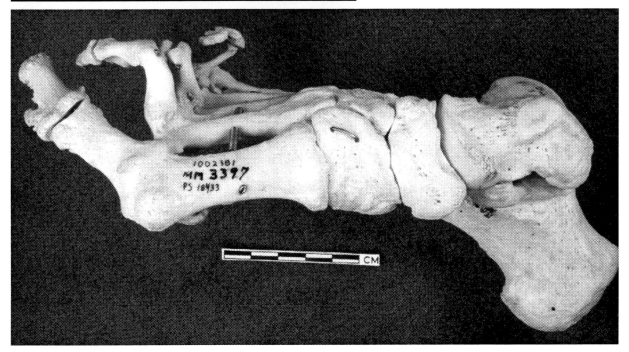

Figure 213. Seronegative spondyloarthrophy of the foot (Mann et al. 1990). (AFIP MM3397)

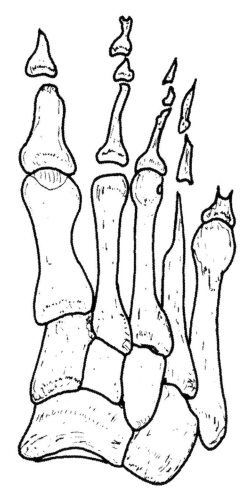

Figure 214. Leprosy of the feet (the talus and calcaneus not shown).

skeletal lesions and paleopathological review; Paterson 1965; Paterson and Job 1964; Patil 2000; Steinbock 1976:192–212; Wastie 1975.

Note severe loss of bone (osteoporosis/osteopenia) and angle of the foot as a result of many years of being wrapped/bound by pulling the forefoot toward the hindfoot (**Figure 215**) This process, practiced by the ancient Chinese from the tenth century until 1911, often resulted in damage to the foot, gangrene, and sometimes death. The gradual process of foot binding, which greatly altered the woman's gait and activities, was begun at a very young age and progressed until the soles of the feet were highly arched (concave), usually by about age three. As a result, the highly decorated shoes ("lotus") for such bound feet were very small (2–3 inches long). Mann et al. 1990. See Ortner 2003:164, Fig. 8–69 for articulated foot deformed by foot binding.

Enlarged nutrient foramina (**Figure 216**) are often seen on adjoin-

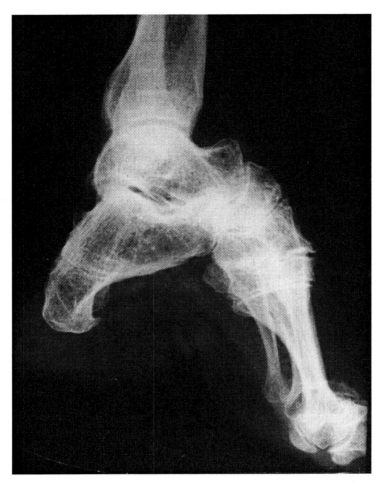

Figure 215. Radiograph of the effects of long-term Chinese foot binding (Mann et al. 1990).

ing metatarsals and may be symmetrical. Fink et al. (1984) reported that normal foramina can attain a size of 1 mm. Although the etiology of enlarged foramina in the foot is unknown, such a condition in the phalanges of the hand has been found in association with thalassemia major (Poznanski 1974), Gaucher's disease (Fink et al. 1984) and leprosy (Møller-Christensen 1967). It is possible that increased blood flow to the extremities is responsible for such enlargement.

Bony ankylosis of any joint (**Figures 212 & 217**) may be the result of many conditions and variables including; infectious disease, developmental or congenital birth defects, endocrinological disturbances or trauma. An accurate interpretation for the most probable cause for the ankylosis must be based on the overall evaluation of the associating skeleton, since it may be a combination of causes to produce the fusion. As an example of indirect causation, Balen and Helms (2001) report ankylosis of the foot as the result of thermal and electrical burns.

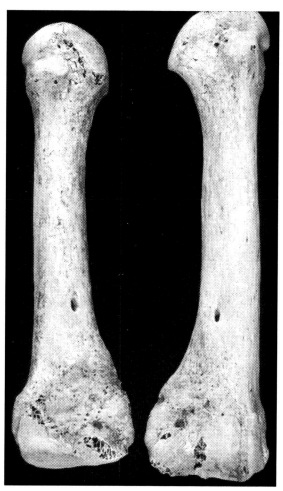

Figure 216. Enlarged nutrient foramina in metatarsals.

Numerous studies have shown that overuse or overstressing activities such as soccer, American football, dancing, and kicking may result in the formation of talotibial osteophytes longer than 1 mm. Tol et al. (2002) investigated hyperplantar flexion in soccer players, measured the magnitude of impact force with the soccer ball, and noted the location of osteophytes in 15 elite soccer players. The authors concluded that ankle impingement syndrome is related to recurrent ball impact, a repetitive microtrauma to the anteromedial aspect of the ankle (Eisele and Sammarco 1993). Other authors have attributed osseous remodeling where the talus abuts with the tibia (**Figure 219**) resulting in the formation of either a squatting facet or osteophyte (along the talus and/or tibia) (Massada 1991). Berberian et al. (2001) found that anterior talotibial osteophytes have a specific pattern of formation and location such that the talar spur (osteophyte) occurs on the medial

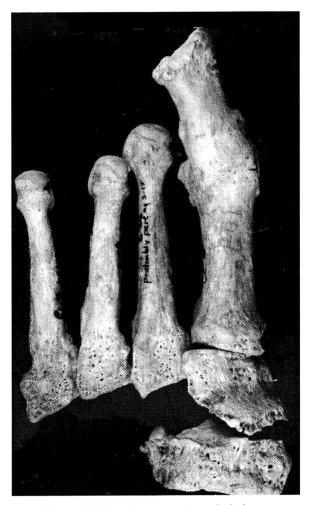

Figure 217. Fused metatarsals and phalanx.

aspect of the talar neck, while the tibial spur forms lateral to the midline. The best way to discern osteophytes along the anterior rim of the tibia and/or talar neck is to articulate the two bones such that they abut, and look for alterations in the two surfaces where they come into contact with one another. An os trigonum may also be present in individuals with footballer's ankle (talotibial osteophytes) (Masciocchi et al. 1998). Haverstock 2001; Hawkins 1988; Lepow and Cafiero 1980; Myers 1987; Parkes et al. 1980; Stoller et al. 1984; Umans 2002; Van Dijk et al. 2002; Vincelette et al. 1972; Zhang et al. 2002.

Tuft resorption (**Figure 220**) in either hand or foot phalanges may reflect psoriatic arthritis (Battistone et al. 1999; Burgener and Kormano 1991). Although rare in most skeletal samples, "spade-like" terminal hand phalanges may be associated with acromegaly (Taylor and Resnick 2000).

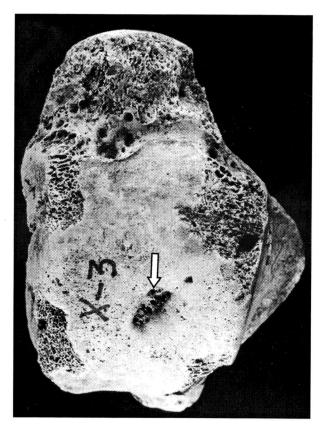

Figure 218. Osteochondritis dissecans of the talar dome (Canale and Belding 1980; Pecina and Bojanic 2004).

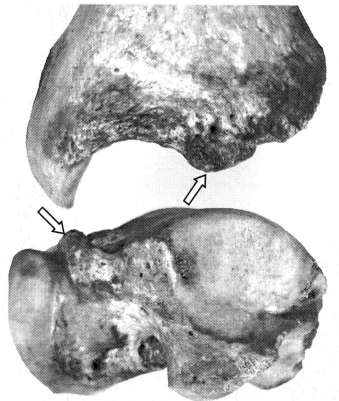

Figure 219. Anterior ankle impingement syndrome resulting in exostoses on the talar neck (arrow) and/or anterior tibia (arrow) (also known as "soccer player's exostosis" or "footballer's ankle").

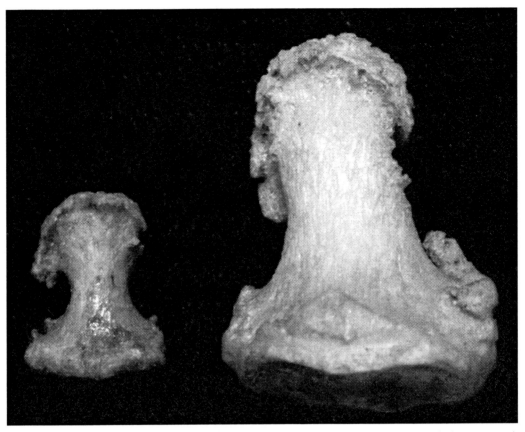

Figure 220. Tufting and osteoarthritis of the distal phalanges (in this case, the foot). (NMNH–H)

If this process (**Figure 221**) is completely separated, the variant is referred to as os trigonum (**Figure 207**). The corresponding irregularly shaped loose bone is commonly overlooked, misidentified, or not recovered with the remains.

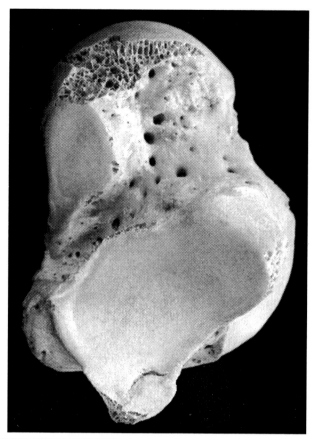

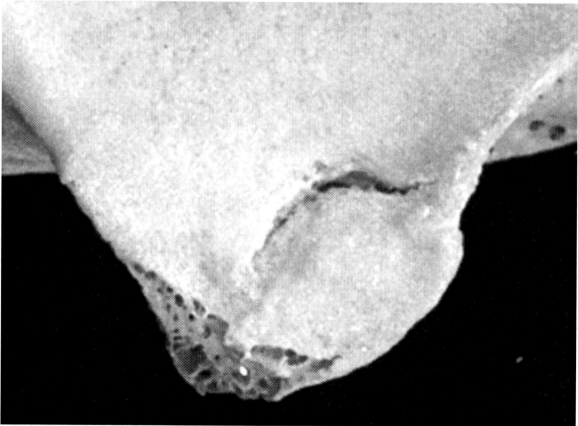

Figure 221a & b. Steida's process of the talus (incomplete separation of the posterior talar process).

Chapter VI

FUNGAL INFECTIONS

Fungi are plant-like organisms that, for the most part, are not harmful to man. In fact, they play an important role in the decay of organic matter. Only a small group of fungi and fungi-like organisms are pathogenic (disease causing) to humans. The crucial factor is the patient's immune response.

Bone lesions in many fungal infections are similar in appearance (e.g. blastomycosis, coccidioidomycosis, and cryptococcosis). Most infections invade the bone by direct extension from a soft tissue lesion. The fungi then burrow into the bone producing cratered or scooped-out lesions that appear as small (0.3 to 1.5 cm), round or ovoid depressions with thin, sharp margins. The lesions tend to be clustered and confluent with minimal periosteal reactive bone. In older lesions the bone may exhibit marked thickening, small spicular bony projections, and shallow craters. Colonna and Gucker 1944; Jones and Martin 1941; Reeves and Pederson 1954.

It is extremely difficult and often impossible to distinguish one form of mycosis from another without soft tissue and a patient's history. Differential diagnosis in mycotic infection depends largely on the patterning of skeletal involvement and fungi associated with the geographic region where the skeleton was recovered (**Figures 222 & Table 1**).

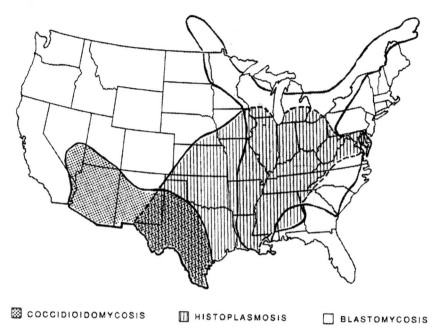

Figure 222. Geographic distribution of fungal infection in the continental United States (from Rippons 1980; Youmans et al. 1980).

Table 1
SELECTED DISEASES THAT PRODUCE BONY REACTIONS
SIMILAR TO MYCOTIC AND ACTINOMYCOTIC INFECTION

Bones Involved	Differential Diagnosis
Vertebrae	Tuberculosis, brucellosis, nonspecific osteomyelitis, metastatic carcinoma
Long bones	Nonspecific osteomyelitis, tuberculosis, treponematosis, various neoplasms
Skull	Treponematosis, tuberculosis, multiple myeloma, metastatic carcinoma, eosinophilic granuloma, histiocytosis X, osteomyelitis

The following are a few of the more commonly encountered fungal organisms or organisms that produce a skeletal response similar to fungi:

Actinomycosis (lumpy jaw, leptothricosis, streptotricosis)–a chronic granulomatous infection caused by the fungal-like ("higher" bacterial) *Actinomyces israelii.* Actinomycosis has a worldwide distribution. There are three anatomical sites of primary infection–cervicofacial (most common), thoracic, and abdominal. The skeletal area affected depends on where the organism enters the body. The lesion will almost always start as a periosteal reaction (periostosis) and new bone formation with eventual erosion of the underlying cortex. Although any bone is susceptible, the mandible and vertebrae are most commonly affected. Actinomycosis of the spine has often been incorrectly identified as tuberculosis (Simpson and McIntosh 1927). However, unlike tuberculosis, actinomycosis frequently results in erosion of the articular facets, laminae, transverse and spinous processes. Goldsand 1989; Ortner and Putschar 1985.

Blastomycosis (North American blastomycosis, Gilchrist's disease, Chicago disease). A chronic granulomatous fungal infection that starts as a pulmonary or cutaneous inflammation that often disseminates to a systemic disease. Blastomycosis is as soil-borne fungus that grows on decaying material and is most commonly found in the southeastern and central United States (Mississippi and Ohio River basins). Males have a higher incidence than females with the average age of onset between 20 and 50 years (Jaffe 1975). Twenty to 50 percent of people with blastomycosis develop skeletal lesions and systemic blastomycosis is fatal 90 percent of the time (Colonna and Gucker 1944; Reeves and Pederson 1954). While all bones are susceptible, the most common sites are the vertebrae, skull, ribs, tibiae, tarsals, and carpals (Jones and Martin 1941). The differential diagnosis between blastomycosis and tuberculosis is that blastomycosis is more diffuse in the skeleton, will often involve several vertebrae and will affect the dorsal as well as the ventral portions. Aufderheide and Rodriguez-Martin 1998:214–215; Colonna and Gucker 1944; Kelley and Eisenberg 1987; Martin and Smith 1939; Moore and Green 1982; Ortner 2003:326; Reeves and Pederson 1954; Zimmerman and Kelley 1982:89.

Coccidioidomycosis (coccidiomycosis, Posadas disease, valley fever, desert rheumatism, San Joaquin fever). Caused by the inhalation of soil-borne fungal spores, this disease is generally a benign inflammation of the respiratory tract. Only a small number of people (usually immune deficient) will contract the disease (Fiese 1958). Coccidioidomycosis is endemic to a well-defined area of southwestern Texas) Mexico, Central and South America (Binford and Conner 1976). All bones are potential sites with the ribs, vertebrae, skull, and bones of the extremities favored (Fiese 1958). Periosteal new bone formation and sequestra are uncommon. The lesions tend to be lytic and favor the cancellous areas of bone. Aufderheide and Rodriguez-Martin 1998:215–217; Bried and Galgiani 1986; Hoeprich 1989; Ortner 2003:326–327; Zeppa et al. 1996; Zimmerman and Kelley 1982:89–90.

Histoplasmosis (Darling's disease, cave disease, Ohio Valley disease, Tingo Maria fever). A common pulmonary infection with worldwide distribution (Binford and Conner 1976). The fungus grows in soil (particularly soil rich in bird feces) and enters the body through the inhalation of spores. Histoplasmosis is endemic in the eastern and central United States. Aufderheide and Rodriguez-Martin 1998:217–218; Ortner 2003:327; Youmans et al. 1980.

For further information refer to Attah and Cerruti 1979; Aufderheide and Rodriguez-Martin 1998:212–222; Brook et al. 1977; Caravio et al. 1977; Dalinka et al. 1971; Daveny and Ross 1969; Echols et al. 1979; Greer 1962; Ortner 2003:325–332; Poswall 1976; Procknow and Loosli 1958; Reeves and Pederson 1954; Seligsohn et al. 1977; Wheat 1989; Zimmerman and Kelley 1989:88–91.

Chapter VII

TREPONEMATOSIS (SYPHILIS)

Treponematosis is a chronic infection caused by a corkscrew-shaped organism (spirochete) of the genus Treponema (meaning "twisted thread"), most commonly called syphilis. Actually, due to the clinical and geographical variation, the infection is divided into four types: pinta, yaws, endemic syphilis, and venereal (and congenital) syphilis (lues or leutic disease) (Table 2). With the variations of the disease and the wide range of its expression in the skeleton, if treponematosis is suspected in an individual, seek the advice of at least one paleopathologist familiar with the disease. Diagnosing bone as syphilitic (or the other forms) is difficult due to the similarity in the skeletal response to the infectious agent as is seen in nonspecific osteomyelitis and other allied infectious diseases. Syphilis is often called the "great imitator" (Thomas 1985). The information about treponematoses offered below should be considered only as an introduction to this variable disease and is not meant to serve as a specific tool for diagnosis, especially in atypical cases.

Pinta (*Treponema carateum*) is the least destructive and most benign form of treponematosis. Pinta (Spanish for "spot," "dot" or "mark") is endemic only to American tropics and is most frequently found in Mexico and parts of Central and South America. Pinta is spread through direct nonsexual contact that initially appears as a skin rash, undergoes pigment changes, and spreads (secondary pinta). The tertiary (final) stage is usually marked by a general depigmentation of the lesions. Pinta is the only treponemal infection that does not cause bone lesions.

Yaws (*Treponema pertenue*), like pinta, is a nonvenereal endemic juvenile disease (Hudson 1940) that is spread through close contact with an infected sore (e.g., children playing). Yaws is found worldwide in hot and humid tropical regions, usually begins in childhood, and starts as a painful localized lesion/sore on an exposed extremity (Binford and Connor 1976). If the sore is located near bone it will induce periostosis (polydactylitis is a frequent finding in early stage yaws Hoeprich 1989). Bone lesions in yaws are rare (ca. 15%) and generally considered indistinguishable from other treponemal infections (Hackett 1963, 1976, 1983). One of the most consistent findings in yaws is anterior bone hypertrophy of the tibia ("boomerang leg"); only rarely does the fibula show this feature. In the late stages, gummatous periostosis and osteomyelitis develop which are almost identical (although not as severe) to tertiary syphilis. Synonyms are bouba, frambesia, tropica, parangi, and pian (Thomas 1985).

Endemic syphilis (treponarid, bejel) is a nonvenereal infection found in warm, arid to semiarid environment (e.g., Africa, the Middle East, and Asia). Endemic syphilis is spread through close contact with infected persons (e.g., kissing) or contaminated objects (e.g., drinking utensils). Binford and Connor (1976) propose that the clinical symptoms of endemic syphilis and yaws are so similar that the infectious agent should be classified as Treponema pertenue (yaws) instead of *T. Pallidum* (venereal and endemic syphilis). Steinbock (1976) suggests that the endemic form is an intermediate manifestation between yaws and venereal syphilis. Hudson (1958) writes ". . . the standard of living and the level of hygiene in a given community determines whether its syphilis will be venereal or non-venereal." Regardless, bone lesions are uncommon and are difficult to distinguish from yaws and venereal syphilis. Current research by Donald J. Ortner suggests the development of joint lesions in some cases of yaws that are not seen in syphilis. (See Ortner 2003:273-319 for a review of the literature, discussion of its evolutionary history, and the paleopathological evidence of treponematosis.)

Venereal syphilis (*Treponema pallidum,* "pale twisted thread") is transmitted through sexual contact. Venereal syphilis is found worldwide and is the only treponemal infection that can be passed from mother to fetus (congenital syphilis). The venereal form, also known as acquired syphilis, uncommonly involves bone (10 to 20%), usually in the tertiary stage of the disease, and two to ten years after the initial infection (Ortner and Putschar 1985; Rudolph 1989; Steinbock 1976). The bones most commonly affected are the tibia and skull, although any bone(s) may be involved (Jaffe 1975).

In the skull (**Figure 223**), syphilis almost always first attacks the outer table, unlike tuberculosis and some neoplasms that start in the diploe and work outwards (**Figures 184, 191 & 192**). The most characteristic cranial lesion is the pattern of scarring seen on the frontal and parietals (occasionally affects the occipital and temporals), called caries sicca (pronounced "sick-uh"). There are three types of lesions that together form caries sicca: stellate scarring, nodes, and cavitations

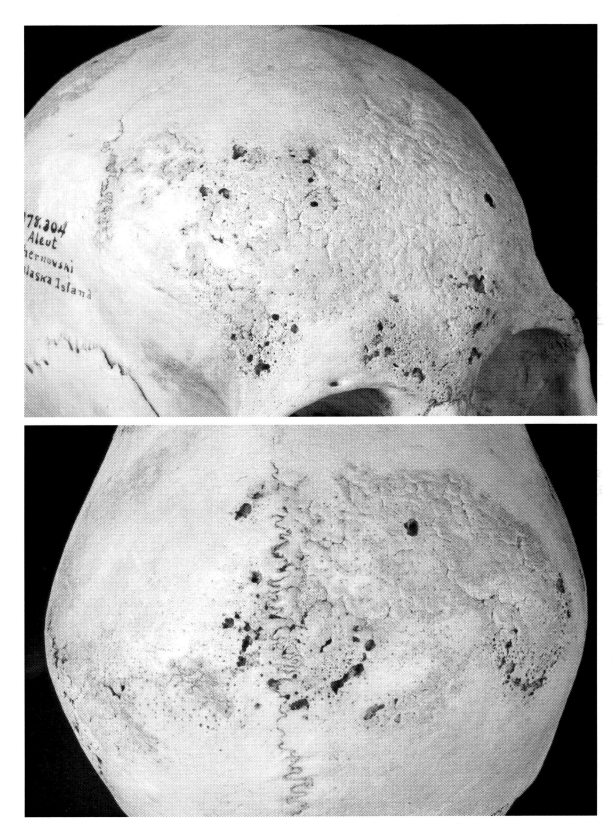

Figure 223a & b. Probable early stage lesions reflecting treponematosis (syphilis?) in the skull of an Alaskan Eskimo.

(Hackett 1976). Cavitations appear as areas of depressed bone that usually do not perforate the inner table. The walls and rims of the depressions are comprised of smooth sclerotic bone (nodes) and stellate (star-like) scars, also known as radial scars. These scars appear as smooth-rimmed furrows that radiate from a central point. Caries sicca represents a healed stage of the disease and will remain visible throughout life. Active areas of early inflammation may be difficult to differentiate from other infections such as tuberculosis, osteomyelitis, neoplasm, and others. See Webb (1995) for treponemal infections in Australian aboriginal populations. (Zimmerman and Kelley 1982:101–102 present an example of tertiary syphilis from the Mutter Museum.)

Table 2
OSSEOUS CHANGES PRESENT IN PATIENTS
WITH LATE CONGENITAL SYPHILIS
(Fiumara and Lessell 1970, 1983)

Stigmata	Boston (1960–1969) n/%	Caribbean (1975–1981) n/%
Hutchinson teeth	171/63.1*	54/100**
Mulberry molars	176/64.9*	71/100**
Frontal bossae	235/86.7	96/96
Short maxillae	227/83.3	100/100
Sternoclavicular thickening	107/39.4	81/81
Relative protuberance of the mandible ("Bull-dog-Jaw")	70/25.8	84/84
High palatal arch	207/76.4	79/79
Saddle nose	199/73.4	92/92
Sabre shin	11/4.1	9/9
Scaphoid scapulae	2/0/7	3/3

*Not a true picture of the incidence because many patients had had their teeth extracted or were edentulous.
**The remainder were edentulous.

For further information refer to Baker and Armelagos 1988 (historical overview); Hackett 1975, 1976; Hudson 1940, 1958; Murray et al. 1956; Ortner 2003:273–319 (for historical and literature as well as paleopathological references); Ortner and Putschar 1985; Stewart and Spoehr 1952; Steinbock 1976:86–169 (for review and paleopathological references); Suzuki 1984; Turk 1995; Zimmerman and Kelley 1982:96–102.

Chapter VIII

TUMORS

A tumor, or neoplasm, is new growth of tissue that is uncoordinated with the normal tissue and may interfere with normal physiology (Tables 3–4) (Madewell et al. 1981a&b). Tumors may be either benign or malignant. Benign tumors are usually slow growing, localized, noninvasive, and usually present no serious threat to the life (e.g., osteoma) (Bushan et al. 1981). Malignant tumors, on the other hand, grow rapidly, spread (metastasize) to other tissue, and often eventually result in death of the individual (e.g., osteosarcoma). Invasion of other tissue by metastasis can occur either through the lymphatic system, bloodstream, or by direct extension of the tumor. Not all tumors fall clearly into these two categories. Giant cell tumors, for example, are generally considered benign but up to 30 percent develop malignant tendencies (Hutter et al. 1962; Madewell et al. 1981a&b).

Tumors are classified according to the tissue of origin or differentiation such as bone or cartilage. The prefix usually indicates the tissue of origin (e.g., osteo = bone, chondro = cartilage). However, not all tumors that affect bone arise from bone and cartilage. For example, liposarcoma (from fat), leiomysarcoma (smooth muscle), fibroma (fibrous tissue), hemangioma (vascular tissue), and neurofibroma (nerve tissue) to name a few. Ly et al. 2003.

It should be remembered that differentiating one type of tumor from another disease condition is extremely difficult (**Figure 224**). This task is made even more difficult when attempting a diagnosis in dry bone without benefit of patient history, soft tissues, or the entire skeleton. For example, was there pain or soft-tissue swelling associated with the tumor? Did the individual sustain trauma at the tumor site? Which bones of the skeleton were affected? All of these considerations must be taken into account if a correct interpretation is to be made. Even with a well-documented patient history an accurate diagnosis is often only narrowed down to a few possibilities that best fit the clinical "picture." If a neoplasm is suspected, first radiograph the bone(s)

225

(refer to Table 3 and 4 and Figure 224, and seek the assistance of a qualified paleopathologist, orthopaedic radiologist, or orthopaedic pathologist for the final interpretation.

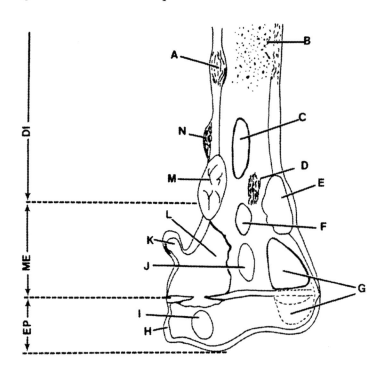

Key:
   A. Cortical fibrous dysplasia, adamantinoma
   B. Round cell lesions–Ewing's Sarcoma, reticulum cell sarcoma, myeloma
   C. Fibrous dysplasia
   D. Fibrosarcoma
   E. Fibroxanthoma (fibrous cortical defect: non-ossifying fibroma)
   F. Bone cyst, osteoblastoma
   G. Giant cell tumor–child: metaphyseal
                adult: "end of bone"
   H. Articular osteochondroma (dysplasia epiphysealis hemimelica)
   I. Chondroblastoma
   J. Enchondroma, chondrosarcoma
   K. Osteochondroma
   L. Osteosarcoma
   M. Chondromyxoid fibroma
   N. Osteoid osteoma
     DI = diaphysis
     ME = metaphysis
     EP = epiphysis

Figure 224. Composite diagram illustrating frequent sites of bone tumors. The diagram depicts the end of a long bone that has been divided into the epiphysis, metaphysis, and diaphysis. The typical sites of common primary bone tumors are labeled. *Adapted from Madewell, J.E. et al.: Radiologic and Pathologic Analysis of Solitary Bone Lesions: Part I Internal Margins. *The Radiologic Clinics of North American, 19*(4), W. B. Saunders Company, Philadelphia, Pennsylvania, 1981.

Table 3
TUMOR AND TUMOR-LIKE CONDITIONS OF BONE BY REGION

	Benign	Malignant	Blastic	Lytic	Location
LONG BONES					
Adamantinoma		X		X	D*
Aneurysmal Bone Cyst	X			X	M
Chondroblastoma	X			X	E/M
Chondrama (enchondroma)	X			X	D/M
Chondrosarcoma		X	X	X	M
Eosinophilic Granuloma	X		X	X	M/D
Ewing's Sarcoma		X	X	X	D/M
Fibrosarcoma		X		X	M
Fribrous Dysplasia	X		X	X	D/M
Fibrous Cortical Defect	X			X	M
Giant Cell Tumor	X			X	E/M
Metastatic Carcinoma		X	X	X	M/D
Multiple Myeloma		X		X	M/D
Non-ossifying Fibroma	X			X	M
Osteoid Osteoma	X		X	X	D
Osteoblastoma	X			X	D/M
Osteochondroma	X		X		M
Osteosarcoma		X	X	X	M
Solitary Bone Cyst	X			X	M
RIB/STERNUM					
Aneurysmal Bone Cyst	X			X	Both
Chondroblastoma	X			X	Ribs
Chondrosarcoma		X	X	X	Both
Fibrous Dysplasia	X		X	X	Ribs
Metastatic Carcinoma		X	X	X	Both
Multiple Myeloma		X		X	Both
Osteoid Osteoma	X		X	X	Ribs
Osteosarcoma		X	X	X	Both
PELVIC/PECTORAL					
Aneurysmal Bone Cyst	X			X	All
Chondroblastoma	X			X	All
Chondrosarcoma		X	X	X	P/S
Metastatic Carcinoma		X	X	X	All
Multiple Myeloma		X		X	P
Osteoid Osteoma	X		X	X	P/S
Osteosarcoma		X	X	X	All
SPINE					
Aneurysmal Bone Cyst	X			X	All
Hemangioma	X			X	T/L
Metastatic Carcinoma		X	X	X	All
Multiple Myeloma		X		X	All
Osteoblastoma	X			X	All

continued

Table 3–*continued*

	Benign	Malignant	Blastic	Lytic	Location
SKULL					
Eosinophilic Granuloma	X		X	X	Calv.
Fibrous Dysplasia	X		X	X	All
Hemangioma	X			X	Calv.
Metastatic Carcinoma		X	X	X	All
Multiple Myeloma		X		X	Calv.
Osteoblastoma	X			X	All
Osteoma	X		X		Calv.
HAND/FOOT					
Chondroblastoma	X			X	Tars.
Chondroma (enchondroma)	X			X	D
Metastatic Carcinoma		X	X	X	All
Osteoblastoma	X			X	All
Osteochondroma	X		X		All
Osteoid Osteoma	X		X	X	All

*Diaphysis; E = epiphysis; M = metaphysis; P = pelvis; S = scapula; T = thoracic; L = lumbar.

For other discussion on clinical and paleopathological cases see: Aufderheide and Rodriguez-Martin 1998:371–392; Brothwell and Sandison 1967:320–345; Coley 1949; Dahlin 1967; Grupe 1988; Hutter et al. 1962; Jaffe 1958; Manchester 1983; Steinbock 1976:316–401; Suzuki 1987; Tkocz and Bierring 1984.

Table 4
TYPICAL AGE RANGE OF THE OCCURRENCE OF BONE TUMORS

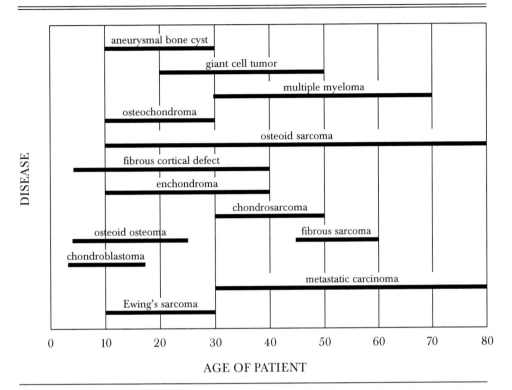

(Reprinted with permission from Griffiths, *Basic Bone Radiology*, Appleton-Century-Crofts, Norwalk, CT, 1981, as modified.)

Chapter IX

PERIMORTEM VERSUS
POSTMORTEM FRACTURES

Forensic pathologists refer to soft tissue injuries usually as antemortem (inflicted during life) (**Figures 227–229**) or postmortem (occurring after death). The degree of tissue reaction and bleeding are the primary indicators. Occasionally wounds fall in between, inflicted while the individual was in the act of dying, and this is termed perimortem. For soft tissues, this period is rather short and precise; in dry bones this term has a slightly different meaning and covers a time that is indistinguishable from antemortem and may persist until weeks after death. The reason for this is twofold. First, many times there is no soft tissue left to aid ante-, peri-, or postmortem determination. Second, living bone is a visco-elastic material with special fracture properties that do not disappear until the elastic (collagen) component deteriorates after death and this may take weeks or months. Thus, in the absence of soft tissue evidence, the perimortem period loses its precision and must include recent antemortem events and an obscure postmortem period during which the bone retains its "fresh" visco-elastic properties.

Perimortem fractures (**Figures 225 & 226**) usually leave sharp, smooth, often beveled fracture lines. Most of the bone at the site of fracture will be present. Fractures of fresh bone often have associated radiating fracture lines at the site of trauma. The fractured ends are usually the same color (as discolored) as the adjacent surface bone (old, dirty appearing). Displaced, curved but still adhering bone fragments or splinters or incomplete fractures with bending of the bone (greenstick fractures), and dirt within the breaks are also associated with perimortem trauma. But it is not always possible to distinguish perimortem fractures from postmortem. Further, because the breakage may have occurred weeks to months after the person's death, but while the bone was still fresh, statements establishing relationship to the cause of death (e.g., gunshot) must be made with caution and judgment. As documented by the authors' research on Civil War soldiers who survived their injuries long enough for remodeling to occur, fracture repair may be evident in as little as 10 days and, in all cases, by two weeks. (Also, Pers. Comm. Paul Sledzik, 1990)

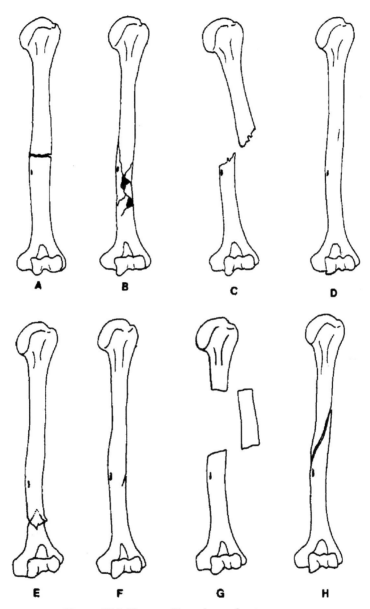

Figure 225. Types of long bone fractures.

Transverse. Fracture in which the bone is broken perpendicular to its long axis (see Greenstick fracture). If the bone is broken in two (A) the fracture is termed complete.

Comminuted. Fracture in which the bone is broken into many pieces or fragments. Common finding in the elderly.

Oblique and displaced. An oblique fracture is one in which the bone is completely broken at an angle diagonal to its long axis (**Figure 229**). Displaced fractures may result in misalignment and overriding of the broken ends. Improper or inadequate immobilization of the fracture may result in a pseudoarthritis or loss in bone length.

Hairline. Minor fracture in which the bone fragments remain in perfect alignment (e.g., "march").

Impacted. Fracturing and subsequent wedging of one bone end into the interior of another. Impaction results in a reduction in the normal length of the bone.

Incomplete. Fracture more severe than a hairline but less severe than a complete with no separation of bone fragments (e.g., sword wound).

Segmental. Fracture in which a significant portion (intact segment) of the bone is displaced. Again, loss of bone or improper immobilization may result in a pseudoarthrosis or loss of bone length.

Spiral. Oblique fracture. Common finding in the elderly and usually resulting from osteoporosis and brittle bones. Spiral fractures are commonly associated with perimortem or "fresh" bone (see **Figure 228**)

Stellate (not shown). Fracture in which lines radiate from the central point of impact (e.g., penetrating wound such as a gunshot; see DiMaio 1999 for information on gunshot wounds) (see **Figure 62**).

Undisplaced (not shown). Fracture in which the bone fragment(s) and host bone remain in approximate anatomical position.

Greenstick (not shown). The thick periosteum surrounding subadult long bones may result in the bending (actually, a slight fracture will be histologically visible) of one side of the cortex and fracture of the other. The appearance is similar to a broken "green stick" (Roser and Clawson 1970; Wilkins 1980).

Figure 226. Perimortem fracture of the fibula resulting in two "butterfly" fractures reflecting the type and direction of force. Note the absence of healing or bone callus (which would indicate that the injury was sustained weeks, months or years before death). (Refer to radiology or orthopaedic texts for further information.)

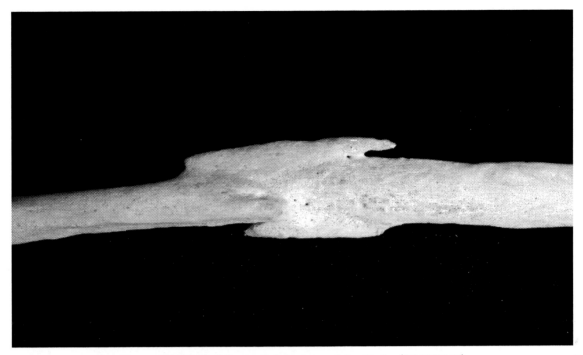

Figure 227. Healed "butterfly" fracture of the fibula. (NMNH–H)

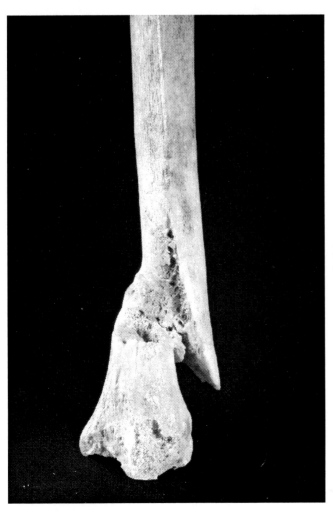

Figure 228. Healing or healed spiral fracture of the distal tibia. (NMNH–H)

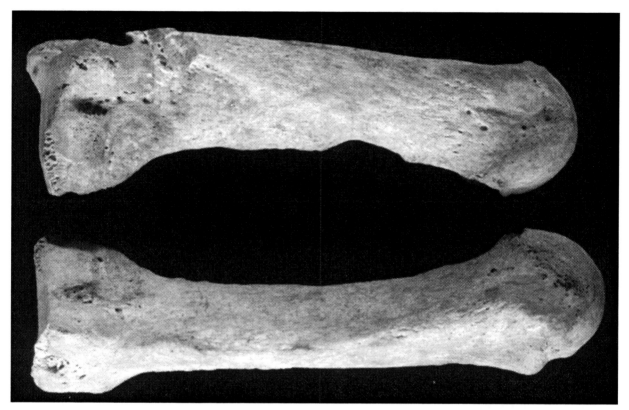

Figure 229. Healed fracture of a metacarpal (top) compared with normal bone.

Although the metacarpal (**Figure 229**) was chosen as an example of a healed displaced long bone fracture, the fracture terminology applies to any tubular bone in the body. In this example, the fracture occurred diagonally at the midshaft and resulted in shortening and overriding of the two broken ends. Evidence of healing can be seen by fusion of the proximal shaft to the distal shaft. Extensive callus formation may accompany a fracture. If the bone appears "swollen" and radiographs reveal large areas of sclerosis (especially if there is a cloaca or draining sinus), it is possible that the fracture site was infected (osteomyelitis) (Resnick and Niwayama 1981). Rockwood and Green 1975; Rogers 1982.

Although there is disagreement as to how long it takes for the earliest signs of periosteal remodeling (callus formation) to become visible on the surface of the cortex, it is probably safe to say that any active remodeling visible in dry bone suggests at least 10 days of healing and, depending on age of the individual, complete remodeling may occur in as little as a few months. Other than nasal bones and vertebrae, healed fractures are relatively uncommon findings in most skeletal samples

Figure 230. Postmortem fractures (more accurately, "breakage").

Postmortem fractures/breakage usually refer to events that occur after death and clearly indicate that the bone has lost its elastic properties (**Figure 230**). Varied conditions of decomposition impact the timing of the loss of bony elasticity. When bone is exposed to the environment for a long period of time it becomes dry, weathered, subject to distortion, breaking, checking, and cracking. Although it is extremely difficult to ascertain what may have caused a particular break, it is often possible to discern postmortem features. Postmortem characteristics reflect its viscous or brittle nature, fracture edges are lighter (sometimes whitish) than surrounding bony areas, irregular to jagged but with blunt/dull edges and little or no beveling, few or no tiny radiating fractures, and small areas of missing bone that become "dust" upon breakage.

See Galloway (1999) and also Wilson (1930) for a more definitive covering of fractures and trauma to the skeleton. See also documentation of Mediaeval battle fractures from the Battle of Towton by Fiorato et al. (2001). Mihran and Tachdjian 1994; Morrissy et al. 2000; Resnick 2002; and Resnick and Niwayama 1988. Also refer to any of the radiology or orthopaedic books and journals such as the *American Journal of Sports Medicine, American Journal of Orthopedics, British Journal of Sports Medicine, Clincial Orthopaedics and Related Research, Journal of Sports Science and Medicine, International Journal of Sports Medicine, Seminars in Musculoskeletal Radiology,* and the *Journal of Bone and Joint Surgery* (American or British) for case reports and types of activity that result in particular fractures, trauma, and rates of healing.

Chapter X

HUMAN SKELETON
(VENTRAL AND DORSAL VIEWS)

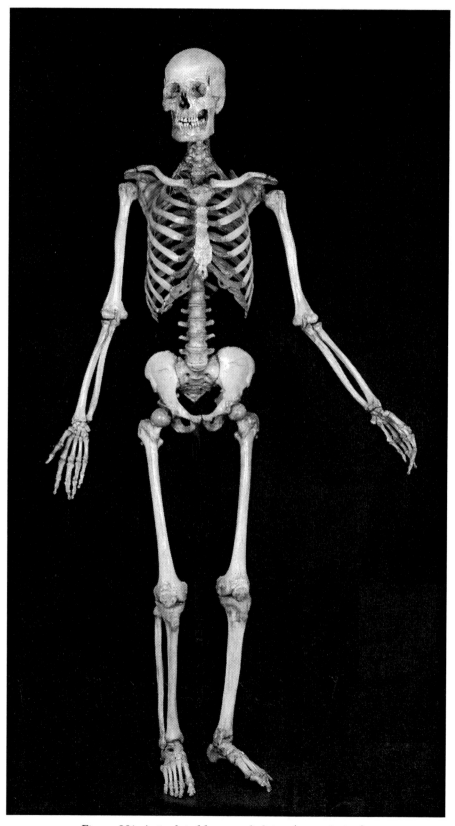

Figure 231. Articulated human skeleton (ventral view).

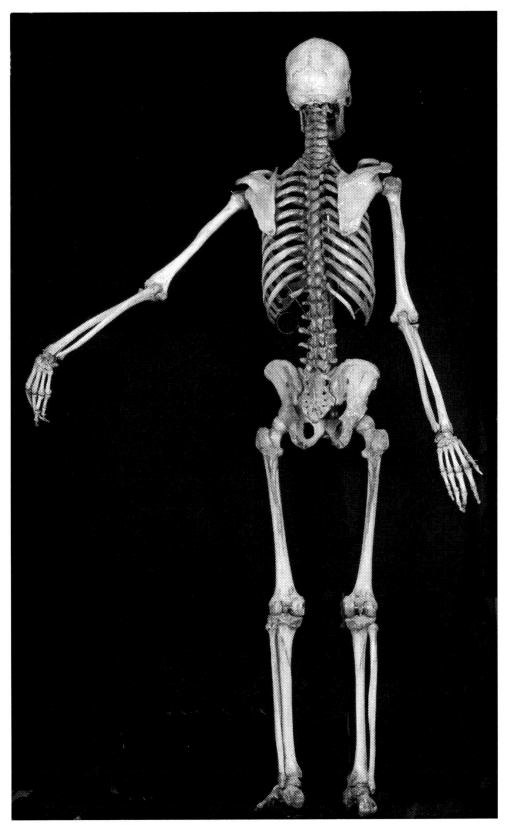

Figure 232. Articulated human skeleton (dorsal view).

Chapter II

MUSCLE ATTACHMENTS

Skull

1. Temporalis
2. Masseter
3. Temporalis
4. Masseter
5. Buccinator

Clavicle

6. Deltoid
7. Pectoralis major
8. Trapezius

Scapula

9. Inferior belly of omohyoid
10. Conoid ligament
11. Trapezoid ligament
12. Pectoralis minor
13. Coracobrachialis and short head of biceps
14. Long head of triceps
15. Subscapularis
16. Serratus anterior

Humerus

17. Supraspinatus
18. Subscapularis
19. Pectoralis major
20. Latissimus dorsi

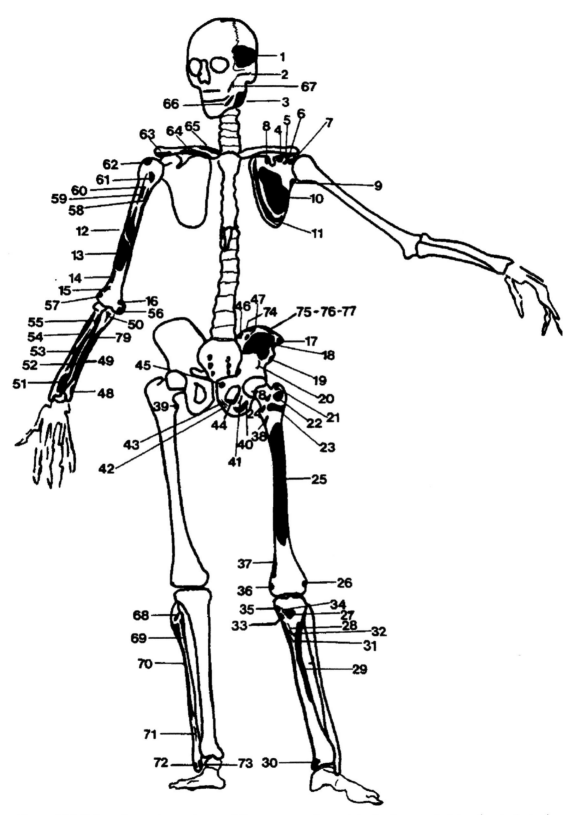

Figure 233. Selected muscle, tendon, and ligament attachments in the human skeleton (ventral view).

21. Teres major
22. Deltoid
23. Brachialis
24. Brachioradialis
25. Extensor carpi radialis longus
26. Pronator teres
27. Common flexor
28. Common extensor

Radius

29. Biceps
30. Supinator
31. Flexor digitorum superficialis, radial head
32. Pronator teres
33. Flexor pollicis longus
34. Pronator quadratus

Ulna

35. Brachialis
36. Flexor digitorum profundus
37. Pronator quadratus

Innominate

38. Erector spinae
39. Iliolumbar ligament
40. Quadratus lumborum
41. Transversus abdominus
42. Internal oblique
43. External oblique
44. Inguinal ligament
45. Sartorius
46. Iliacus
47. Rectus femoris
48. Psoas minor
49. Adductor longus
50. Adductor brevis
51. Gracilis
52. Obturator externus
53. Quadratus femoris

54. Adductor magnus

Femur

55. Piriformis
56. Gluteus minimus
57. Iliofemoral ligament
58. Vastus lateralis
59. Iliofemoral ligament
60. Vastus medialis
61. Vastus intermedius
62. Adductor magnus
63. Fibular collateral ligament
64. Tibial collateral ligament

Tibia

65. Semimembranosus
66. Vastus medialis
67. Patellar ligament
68. Tibial collateral ligament
69. Sartorius
70. Gracilis
71. Semitendinous
72. Tibialis anterior
73. Medial collateral ligament

Fibula

74. Peroneus longus
75. Extensor digitorum longus
76. Peroneus longus
77. Peroneus tertius
78. Calcaneofibular ligament
79. Anterior talofibular ligament

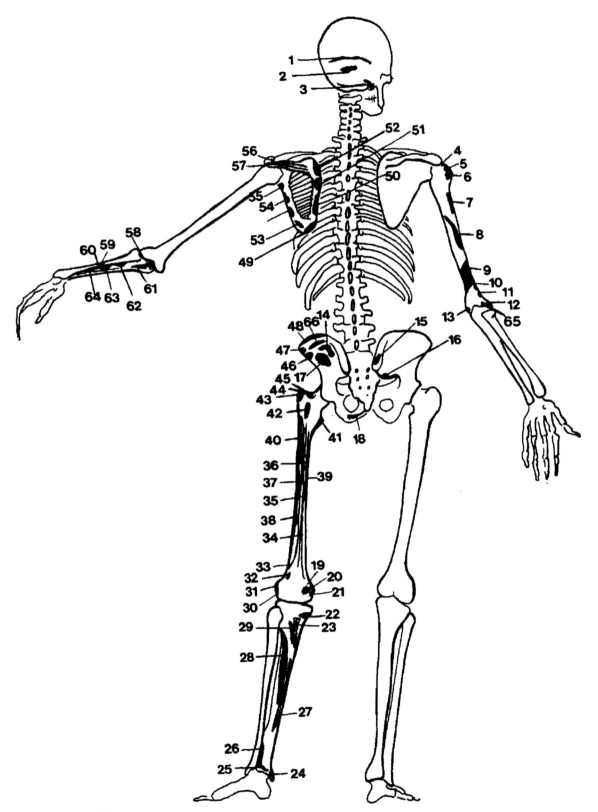

Figure 234. Selected muscle, tendon, and ligament attachments in the human skeleton (dorsal view).

Skull

1. Trapezius
2. Semispinalis capitis
3. Sternocleidomastoid

Scapula

4. Levator scapulae
5. Rhomboid minor
6. Rhomboid major
7. Trapezius
8. Deltoid
9. Long head of triceps
10. Teres minor
11. Teres major
12. Latissimus dorsi

Humerus

13. Supraspinatus
14. Infraspinatus
15. Teres minor
16. Lateral head of triceps
17. Medial head of triceps
18. Medial head of triceps
19. Brachioradialis
20. Extensor carpi radialis longus
21. Common extensor
22. Anconeus
23. Common flexor

Radius

24. Extensor pollicis longus
25. Adductor pollicis longus

Ulna

26. Triceps
27. Anconeus
28. Pronator teres

29. Apponeurotic attachment of Flexor digitorum profundus, Flexor carpi ulnaris and Extensor carpi ulnaris
30. Extensor indicis

Innominate

31. Gluteus minimus
32. Tensor fasciae latae
33. External oblique
34. Inguinal ligament
35. Sartorius
36. Gluteus minimus
37. Adductor magnus
38. Gluteus maximus
39. Piriformis

Femur

40. Obturator internus
41. Obturator externus
42. Gluteus medius
43. Quadratus femoris
44. Psoas and iliacus
45. Gluteus maximus
46. Adductor brevis
47. Vastus intermedius
48. Vastus medialis
49. Vastus lateralis
50. Adductor longus
51. Adductor magnus
52. Short head of biceps
53. Plantaris
54. Lateral head of gastrocnemius
55. Fibular collateral ligament
56. Medial head of gastrocnemius
57. Adductor magnus
58. Tibial collateral ligament

Tibia

59. Semimembranous
60. Popliteus

61. Soleus
62. Tibialis posterior
63. Flexor digitorum longus
64. Interosseous ligament
65. Posterior tibiofibular ligament/inferior transverse ligament
66. Medial collateral ligament

BIBLIOGRAPHY

Abraham, E., Verinder, D. G. R., and Sharrard, W. J. W.: The treatment of flexion contracture of the knee in myelomeningocele. *Journal of Bone and Joint Surgery 59-B*(4):433–438, 1997.

Adams, J. D., and Leonard, D. R.: A developmental anomaly of the patella frequently diagnosed as fracture. *Surgery, Gynecology, Obstetrics 41:*601–604, 1925.

Adams, I. D.: Osteoarthrosis and sport. *Journal of the Royal Society of Medicine 72*(3):185–187, 1979.

Adams, J. L.: The supracondyloid variation in the human embryo. *Anatomical Record 59:* 315, 1934.

Adelaar, R. S. (Ed.).: *Complex foot and ankle trauma.* Philadelphia, Lippincott-Raven, 1999.

Ait-Ameur, A., Wakim, A., Dubousset, J., Kalifa, G., and Adamsbaum, C.: The AP diameter of the pelvis: A new criterion for continence in the exstrophy complex? *Pediatric Radiology 31*(9):640–645, 2001.

Akrawi, F.: Is bejel syphilis? *British Journal of Venereal Disease 26:* 115–123, 1949.

Albright, F., Smith, P.A., and Richardson, A.M.: Postmenopausal osteoporosis: Its clinical features. *Journal of American Medical Association 116:*2465, 1941.

Alexander, C. J.: Osteoarthritis: a review of old myths and current concepts. *Skeletal Radiology 19:*327–333, 1990.

Allbrook, D. B.: The East African vertebral column. A study in racial variability. *American Journal of Physical Anthropology 13:*489–513, 1955.

Allen, H.: *A System of Human Anatomy, Sec. II, Bones and Joints.* Philadelphia, 1882.

Allison, M.J., and Gerszten, E.: *Paleopathology in South American Mummies.* Richmond:Virginia Commonwealth University, 1982.

Allison, M.J., Gerszten, E., Munizaga, J., Santoro, C.: La practica de la formacion craneana entre los pueblos andinos precolumbinos. *Chungara 8:*238–260, 1981.

Altchek, M.: Congenital clubfeet (letters to the editor). *Clinical Orthopaedics and Related Research 130:* 303–305, 1978.

Anand, V. T., Latif, M. A., and Smith, W. P.: Defects of the external auditory canal: a new reconstruction technique. *Journal of Laryngology Otology 114*(4):279–282, 2000.

Andersen, J. G.: *Studies on the Medieval diagnosis of leprosy in Denmark. An osteoarchaeological, historical and clinical study.* Copenhagen: Costers Bogtrykkeri, 1969.

Andersen, J. G.: *The Medieval diagnosis of leprosy.* In: Ortner D. J., and Aufderheide, A. C, (Eds.). Human palaeopathology: Current syntheses and future options. Washington, DC: Smithsonian Institution Press, pp. 205–208, 1991.

Anderson, J.E.: The people of Fairty. *National Museum of Canada Bulletin 193:* 128–129, 1963.

____: The osteological diagnosis of leprosy. Proceedings of the Paleopathology Association 4th European Meeting, Middleberg, Antwerpen, 1982.

Anderson, T.: Calcaneus secundarius: An osteo-archaeological note. *American Journal of Physical Anthropology 77:*529–531, 1988.

Angel, J. L.: Osteoporosis: Thalassemia? *American Journal of Physical Anthropology 22:*369–373, 1964.

____: Porotic hyperostosis, anemias, malarias and marshes in prehistoric eastern Mediterranean. *Science 153:*760–763, 1966.

___: Colonial to modern skeletal change in the USA. *American Journal of Physical Anthropology* *45:*723–735, 1976.

___: History and development of paleopathology. *American Journal of Physical Anthropology* *56:*509–515, 1981.

Angel, J. L., Kelley, J.O., Parrington, M., and Pinter, S.: Life stresses of the free black community as represented by First African Baptist Church, Philadelphia. *American Journal of Physical Anthropology 74:*213–229, 1987.

Antoniades, D. Z., Belazi, M., and Papanayiotou, P.: Concurrence of torus palatinus with palatal and buccal exostoses: Case report and review of the literature. *Oral Surgery Oral Medicine Oral Pathology 85*(5):552–557, 1998.

Apley, A. G. and Solomon, L.: *Concise system of orthopaedics and fractures.* London: Butterworth and Company, 1988.

Arey, L. B.: *Developmental anatomy: A textbook and laboratory manual of embryology,* 6th ed. Philadelphia, E. B. Saunders, 1966.

Arnay-de-la-Rosa, M., Velasco-Vazquez, J., Gonzalez-Reimers, E., and Santolaria-Fernandez F.: Auricular exostoses among the prehistoric population of different islands of the Canary archipelago. *Annals Otology Rhinology Laryngology 110*(11):1080–1083, 2001.

Aronica-Pollak, P. A., Stefan, V. H., and Mclemore, J.: Coronal cleft vertebra initially suspected as an abusive fracture in an infant. *Journal of Forensic Sciences 48*(4):836–838, 2003.

Arriaza, B. T., Salo, W., Aufderheide, A. C., and Holcomb, T. A.: Pre-Columbian tuberculosis in Northern Chile: Molecular and skeletal evidence. *American Journal Physical Anthropology 98:*37–45, 1995.

Ashley, G.T.: The relationship between the pattern of ossification and the definitive shape of the mesosternum in Man. *Journal of Anatomy 90:*87–105, 1956.

Attah, C. A.: and Cerruti, M. M.: Aspergillus osteomyelitis of the sternum after cardiac surgery. *New York State Journal of Medicine 79:*1420–1421, 1979.

Aufderheide, A. C., and Rodriguez-Martin, C.: *The Cambridge Encyclopaedia of Human Paleopathology.* Cambridge University Press, 1998.

Baker, B. J., and Armelagos, G. J.: The origin and antiquity of syphilis. *Current Anthropology 29*(5):703–737, 1988.

Balen, P. F., and Helms, C. A.: Bony ankylosis following thermal and electrical injury. *Skeletal Radiology 30:*393–397, 2001.

Banadda, B. M., Gona, O., Vaz, R., and Ndlovu, D. M.: Calcaneal spurs in a black African population. *Foot and Ankle 13*(6):352–354, 1992.

Barnard, L. B., and McCoy, S. M.: The supracondyloid process of the humerus. *Journal of Bone and Joint Surgery 28:*845, 1946.

Barnes, E.: *Developmental defects of the axial skeleton in paleopathology.* University of Oklahoma Press, Niwot, 1994.

Barnett, C.H.: Squatting facets on the European talus. *Journal of Anatomy 88*(4):509–513, 1954.

Barrie, H. J.: Osteochondritis dissecans 1887–1987: A centennial look at Konig's memorable phrase. *Journal of Bone and Joint Surgery 69–B*(5):693–695, 1987.

Barry, H. C.: *Paget's disease of bone.* London: E&S Livingstone Ltd., 1969.

Bass, W. M.: *Human osteology: A laboratory and field manual,* 4th ed. Columbia, Missouri Archaeological Society Special Publication #2, 1995.

Bassiouni, M.: Incidence of calcaneal spurs in osteoarthrosis and rheumatoid arthritis, and in control patients. *Annals of the Rheumatic Diseases 24:*490–493, 1965.

Battistone, M. J., Manaster, B. J., Reda, D. J., and Clegg, D. O.: The prevalence of sacroiliitis in psoriatic arthritis: New perspectives from a large, multicenter cohort. *Skeletal Radiology 28:*196–201, 1999.

Bayliss, M.: Responses in human cartilage in relation to age. In Russell R, G. G., and Dieppe, P. (Eds.) *Osteoarthritis: Current research and prospects for pharmacological intervention.* London:IBC, 1991.

Beckly, D. E., Anderson, P. W., and Pedegana, L. R.: The Radiology of the Subtalar Joint with Special Reference to Talo-Calcaneal Coalition. *Clinical Radiology 26:*333–341, 1975.

Beighton, P.: *Inherited Disorders of the Skeleton.* New York: Churchill Livingstone, 1978.

Bender, P. l.: Genetics of cleft lip and palate. *Journal Pediatric Nursing 15*(4):242–249, 2000.

Benjamin, M., Rufai, A., and Ralphs, J. R.: The mechanism of formation of bony spurs (enthesophytes) in the achilles tendon. *Arthritis Rheumatism 43*(3):576–583, 2000.

Benzel, C. E.: *Biomechanics of spine stabilization.* American Association of Neurological Surgeons, Illinois, 2001.

Berberian, W. S., Hecht, P. J., Wapner, K. L., and DiVerniero, R.: Morphology of Tibiotalar osteophytes in anterior ankle impingement. *Foot and Ankle International 22*(4):313–317, 2001.

Bergman, R. A., Thompson, S. A., Afifi, A. K., and Saadeh, F. A.: *Compendium of human anatomic variation: Text, atlas, and world literature.* Baltimore, Urban and Schwarzenberg, 1988.

Bergsma, D. (Ed.): *Birth Defects Conpendium.* 2nd Ed. New York:Alan R. Liss, 1978.

Bernhard, J.: [Seronegative spondyloarthropathies] [article in German]. *Ther Umsch 59*(10): 529–534, 2002.

Berry, A. C.: Factors affecting the incidence of non-metrical skeletal variants. *Journal of Anatomy 120*:519–535, 1975.

Berry, A. C., and Berry, R. J.: Epigenetic variation in the human cranium. *Journal of Anatomy 101*(2):361–379, 1967.

Berryman, H. E., and Gunther, W. M.: Keyhole Defect Production in Tubular Bone. *Journal of Forensic Sciences 45*(2):483–487, 2000.

Berryman, H. E., and Symes, S. A.: Recognizing gunshot and blunt cranial trauma through fracture interpretation. In Reichs, K. J. (Editor). *Forensic Osteology: Advances in the identification of human remains,* 2nd edition. Springfield, Charles C Thomas, 1998.

Bertaux, T. A.: *L'humerus et le femur* [article in French]. Lile, France, 1891.

Bhalaik, V., Chhabra, S., Walsh, H. P. J.: Bilateral coexistent calcaneonavicular and talocalcaneal tarsal coalition: A case report. *The Journal of Foot and Ankle Surgery 41*(2): 129–134, 2002.

Binford, C. H., and Conner, D. H.: *Pathology of tropical and extraordinary diseases.* Washington, D.C., Armed Forces Institute of Pathology, Vol.1, 1976.

Blackburne, J. S., and Velikas, E. P.: Spondylolisthesis in children and adolescents. *Journal of Bone and Joint Surgery 59-B*(4):490–494, 1977.

Blackwood, H. J. J.: Arthritis of the mandibular joint. *British Dental Journal 115*(8):317, 1963.

Blakey, M. L., Leslie, T. E., and Reidy, J. P.: Frequency and chronological distribution of dental enamel hypoplasia in enslaved African American: A test of the weaning hypothesis. *American Journal of Physical Anthropology 95*:371–383, 1994.

Bland, J. H. *Disorders of the cervical spine: Diagnosis and medical management* (2nd Ed.). Philadelphia: W. B. Saunders, 1994.

Bloch, I.: *History of syphilis, system of syphilis.* London, Hodder & Stoughton, 1908.

Bluestone, C. D., and Klein, J. O.: *Otitis media in infants and children.* Philadelphia, Saunders, 1988.

Bohne, W. H. O.: Tarsal coalition. *Current Options in Pediatrics 13*:29–35, 2001.

Boldsen, J. L.: Epidemiological Approach to the Paleopathological Diagnosis of Leprosy. *American Journal of Physical Anthropology 115*:380–387, 2001.

Bolm-Audroff, U.: Intervertebral disc disorders due to lifting and carrying heay weights. *Medical Orthopedic Technology 112*:293–296, 1992.

Borenstein, D. G., Wiesel, S. W., and Boden, S. D.: *Low back pain: Medical diagnosis and comprehensive management* (2nd ed.). W. B. Saunders: Philadelphia, 1995.

Boyer, M. I., and Gelberman, R. H.: Operative correction of swan-neck and boutonniere deformities in the rheumatoid hand. *Journal American Academy Orthopedic Surgery 7*(2):92–100, 1999.

Boulle, E. L.: Evolution of two human skeletal markers of the squatting position: A diachronic study from antiquity to the modern age. *American Journal of Physical Anthropology 116*(1):50–56, 2001.

Boyer, G. S., Templin, D. W., Cornoni-Huntley, J. C., Everett, D. F., Lawrence, R. C., Heyse,

S. F., Miller, M. M., and Goring, W. P.: Prevalence of spondylarthropathies in Alaskan Eskimos. *Journal of Rheumatology 21:*2292–2297, 1994.

Bowen, J. R., Kumar, V. P., Joyce, John J. III, and Bowen, J. C.: Osteochondritis dissecans following Perthes' disease. In Burwell, R. B. and Harrison, M. H. M. (Eds.): *Clinical Orthopaedics and Related Research 209:*49–56, 1986.

Bradford, D. S.: Spondylolysis and spondylolisthesis. In Chou, S. N. and Seljeskog, E. L (Eds.): *Spinal deformities and neurological dysfunction.* New York: Raven Press, 1978.

Bradley, J., and Dandy, D. J.: Osteochondritis dissecans and other lesions of the femoral condyles. *Journal of Bone and Joint Surgery 71–B*(3);518–522, 1989.

Brannon, E. W.: Cervical rib syndrome: An analysis of nineteen cases and twenty-four operations. *Journal of Bone and Joint Surgery 45–A*(5):977–998, 1963.

Bridges, P.S.: Spondylolysis and its relationship to degenerative joint disease in the prehistoric Southeastern United States. *American Journal of Physical Anthropology 79:*321–329, 1989.

Bried, M.J., and Galgiani, J. N.: Coccidiodides immitis infections in bones and joints. *Clinical Orthopaedics and Related Research 211:*235–243, 1986.

Brook, C. J., Ravikrishnan, K. P., and Weg, J. G.: Pulmonary and articular sporotricosis. *American Review of Respiratory Disorders 116:*141–143, 1977.

Brothwell, D. R.: *Digging up bones: The excavation treatment and study of human skeletal remains.* Ithaca: Cornell University Press, 1981.

____: The real history of syphilis. *Science Journal 6:*27–32, 1970.

____: Biparietal thinning in early Britian. In Brothwell, D. R. and Sandison, A.T (Eds): *Disease in antiquity. A survey of the diseases injuries and surgery of early populations.* Springfield: C. C Thomas, 1967.

Brothwell, D. R., and Sandison, A. T. (Eds.): *Diseases in antiquity. A survey of the diseases injuries and surgery of early populations.* Springfield: Thomas, 1967.

Brower, A. C.: Cortical defect of the humerus at the insertion of the pectoralis major. *American Journal of Roentgenology 128:*677–678, 1977.

Brower, A. C., and Allman, R. M.: The neuropathic joint: A neurovascular bone disorder. *Radiologic Clinics of North America 19*(4):571–580, 1981.

Brown, G. P., Feehery, R. V., Jr., and Grant, S. M.: Case study: The painful os trigonum syndrome. *Journal of Orthopaedics Sports and Physical Therapy 22*(1):22–25, 1995.

Browner, B. D., Jupiter, J. B., Levine, A. M., and Trafton, P. G.: Volume 2, 3rd edition. Philadelphia: Saunders, 2003.

Brukner, P., Bradshaw, C., Kahn, K. M., White, S., and Crossley, K.: Stress fractures: A review of 180 cases. *Clinical Journal of Sport Medicine 6*(2):85–89, 1996.

Bryceson, A., and Pfaltzgraff, R. E.: *Leprosy* (3rd ed.). Edinburgh: Churchill Livingstone, 1990.

Buikstra, J. E., and Williams S.: *Tuberculosis in the Americas: Current perspectives.* In Ortner, D. J., and Aufderheide, A.C. (Eds.). Human Paleopathology: current syntheses and future options. Washington, D.C.: Smithsonian Institution Press, pp. 161–172, 1991.

Bullough, P. G.: Ivory exostoses of the skull. *Postgraduate Medical Journal 41:*277–281, 1965.

Bulos, S.: Herniated intervertebral lumbar disc in the teenager. *Journal of Bone and Joint Surgery 55–B:*273–278, 1973.

Bunnell, W. P.: Back pain in children. *Orthopedic Clinics of North America 13*(3):587–604, 1982.

Burgener, F. A., and Kormano, M.: *Differential Diagnosis in Conventional Radiology* (2nd ed.). New York: Georg Thieme Verlag Stuttgart, 1991.

Burke, M. J., Fear, E. C., and Wright, V.: Bone and joint changes in pneumatic drillers. *Annals Rheumatic Diseases 36:*276–279, 1977.

Burman, M. S., and Lapidus, P. W.: Functional disturbances caused by inconstant bones and sesamoids of the foot. *Archives of Surgery 22:*936, 1931.

Burry, H. C.: Sport exercise and arthritis. *British Journal Rheumatology 25:*386–388, 1987.

Busch, M. T., and Morrissy, R. T.: Slipped capital femoral epiphysis. *Orthopaedic Clinics of North America 18:*637–647, 1987.

Bushan, B., Watal, G., Ahmed, A., Saxena, R., Goswami, K., Pathania, A. G.: Giant ivory osteoma of frontal sinus. *Australasian Radiology 31*(3):306–308, 1987.

Byers, S.: Technical note: Calculation of age at formation of radiopaque transverse lines. *American Journal of Physical Anthropology 85*(3):339–343, 1991.

Cady, L. D.: The incidence of the supracondyloid process in the insane. *American Journal of Physical Anthropology 5:*35–49, 1921.

Caffey, J.: *Pediatric x-ray diagnosis,* 6th ed. Chicago, Year Book Medical Publishers, 1972.

Calin, A.: Ankylosing spondylitis. *Clinics in Rheumatic Diseases 11*(1):41–60, 1985.

Callahan, J. J.: Interesting notes on bipartite patellae. *U. S. Naval Medical Bulletin 48:* 229–233, 1948.

Camarda, A. J., Deschamps, C., and Forest, D.: II. Stylohyoid chain ossification: A discussion of etiology. *Oral Surgery Oral Medicine Oral Pathology 67*(5):515–520, 1989.

Campillo, D.: Herniated intervertebral lumbar discs in an individual from the Roman Era, exhumed from the "Quinta de San Rafael" (Tarragona, Spain). *Journal of Paleopathology 2*(2):89–94, 1989.

Canale, G., Scarsi, M., and Mastragostino, S.: Hip deformity and dislocation in spina bifida. *Italian Journal of Orthopaedics and Traumatology 18*(2):155–165, 1992.

Canale, S. T., and Belding, R. H.: Osteochondral lesions of the talus. *Journal of Bone and Joint Surgery 62-A*(1):97–102, 1980.

Canale, S. T., Griffin, D. W., and Hubbard, C. N.: Congenital muscular torticollis. A long term follow-up. *Journal of Bone and Joint Surgery 64:*810–816, 1982.

Caraveo, J., Trowbridge, A. A., Amaral, B. W., Green, J. B., Cain, P. T., and Hurley, D. L.: Bone marrow necrosis associated with a mucor infection. *American Journal of Medicine 62:*404–408, 1977.

Carlson, D., Armelagos, G., and Van Gerven, D.: Factors influencing the etiology of cribra orbitalia in prehistoric Nubia. *Journal of Evolution 3:*405–410, 1974.

Carney, C. N., and Wilson, F. C.: Infections of bones and joints. In Wilson, F.C (Ed.): *The Musculoskeletal System.* Philadelphia. Lippincott, 1975.

Carpintero-Benitez, P., Logrono, C., Collantes-Estevez, E.: Enthesopathy in leprosy. *Journal Rheumatology 23*(6):1020–1021, 1996.

Casscells, S. W.: The arthroscope in the diagnosis of disorders of the patellofemoral joint. *Clinical Orthopaedics and Related Research 144:*45–50, 1979.

Caterall, A.: *Legg-Calve-Perthes Disease.* New York: Churchill Livingstone, 1982.

Cave, A. J. E.: The earliet English example of bilateral cervical rib. *British Journal of Surgery 29*(113):47–51, 1941.

____: The nature and morphology of the costoclavicular ligament. *Journal of Anatomy 95:*170–179, 1961.

Cecil, R.C. and Loeb, R.F. *A textbook of medicine.* 8th edition. Philadelphia:W.B. Saunders, 1951.

Chi, J., and Harkness, M.: Elongated stylohyoid process: A report of three cases. *New Zealand Dental Journal 95*(419):11–13, 1999.

Chimenos-Kustner, E., Batle-Trave, I., Velasquez-Rengijo, S., Garcia-Carabano, T., Vinals-Iglesias, H., and Rosello-Llabres, X.: Appearance and culture: Oral pathology associated with certain "fashions" (Tattos, piercings, etc.). *Medical Oral 8*(3):197–206, 2003.

Chohayeb, A. A., and Volpe, A. R.: Occurrence of torus palatinus and mandibularis among women of different ethnic groups. *American Journal of Dentistry 14*(5):278–280, 2001.

Choi, S., and Harris, L.: Aortic nonanastomotic pseudoaneurysm eroding lumbar vertebra– A case report. *Vascular Surgery 35*(3):245–250, 2001.

Chopra, S. R. K.: The cranial suture closure in monkeys. *Proceedings of the Zoological Society of London 128:*67, 1957.

Chung, S. M. K., and Nissenbaum, M. M.: Congenital and developmental defects of the shoulder. *Orthopedic Clinics of North America 6:*381–392, 1975.

Churchill, S. E., and Morris, A. G.: Muscle marking morphology and labour intensity in pre-historic Khoisan foragers. *International Journal of Osteoarchaeology 8:* 390–411, 1998.

Cimen, M., and Elden, H.: Numerical variations in human vertebral column: A case report. *Okajimas Folia Anatomy Japan 75:*297–303, 1999.

Claffey, T. J.: Avascular necrosis of the femoral head: An anatomical study. *Journal of Bone and Joint Surgery,* 42–B:802–809, 1960.

Clanton, T. O., and DeLee, J. C.: Osteochondritis dissecans: History, pathophysiology and current treatment concepts. *Clinical Orthopaedics and Related Research* 167:50–64, 1982.

Cockburn, T. A.: The origin of the treponematoses. *Bulletin World Health Organization* 24:221–228, 1961.

Cockburn, A., and Cockburn, E.: *Mummies, Disease and Ancient Cultures* (abridged edition). Cambridge: Cambridge University Press, 1980.

Cockshott, P. W.: Anatomical anomalies observed in radiographs of Nigerians - (I) Thoracic. *West African Medical Journal* 7(4):179, 1958.

Cohen, M. M., Jr.: History, terminology, and classification of craniosynostosis. In Cohen, M. M., Jr. (Ed.): *Craniosynostosis: Diagnosis, evaluation, and management.* New York, Raven, 1986.

Cohen, M M., Jr. and MacLean, R. E.: *Craniosynostosis: Diagnosis, evaluation, and management* (2nd ed.). Oxford University Press, 2000.

Coleman, J. R., Jr., and Sykes, J. M.: The embryology, classification, epidemiology, and genetics of facial clefting. *Facial Plastic Surgery Clinics of North America* 9(1):1–13, 2001.

Coley, B.: *Neoplasms of bone and related conditions.* 2nd Ed. New York, Paul B. Hoeber, 1950.

Collet, S., Bertrand, B., Cornu, S., Eloy, P., and Rombaux, P.: Is septal deviation a risk factor for chronic sinusitis? Review of literature. *Acta Oto-rhino-laryngologica Belgica* 55(4):299–304, 2001.

Collier, B.D, Johnson, R.P., Carrera, G.F., Meyer, G.F., Schwab, J.P., Flatley, T.J., Isitman, A.T., Hellman, R.S., Zielonka, J.S. and Knobel, J.: Painful spondylolysis or spondylolisthesis studied by radiography and single-photon emission computed tomography. *Radiology* 154:207–211, 1985.

Colonna, P. C. and Gucker, T.: Blastomycosis of the skeletal system. *Journal of Bone and Joint Surgery* 26:322–328, 1944.

Colquhoun, J.: Butterfly vertebra or sagittal cleft vertebra. *American Journal of Orthopaedic Surgery* 10(2):44–50, 1968.

Congdon, R. T.: Spondylolisthesis and vertebral anomalies in skeletons of American aborigines. *Journal of Bone and Joint Surgery* 14–B:511–524, 1931.

Conner, A. N.: The treatment of flexion contractures of the knee in poliomyelitis. *Journal of Bone and Joint Surgery* 52–B(1):138–144, 1970.

Cook, D. C.: Subsistence base and health in prehistoric Illinois Valley: Evidence from the human skeleton. *Medical Anthropology* 4:109–124, 1979.

Cook, D. C., and Buikstra, J. E.: Health and differential survival in prehistoric populations: Prenatal dental defects. *American Journal of Physical Anthropology* 51:649–664, 1979.

Coons, M. S., and Green, S. M.: Boutonniere deformity. *Hand Clinics* 11(3):387–402, 1995.

Cooper, P.D., Stewart, J.H. and McCormick, M.S.: Development and morphology of the sternal foramen. *American Journal of Forensic Medicine and Pathology* 9(4):342–347, 1988.

Correll, R. W., Jensen, J. L., and Rhyne, R. R.: Lingual cortical mandibular defects: A radiographic incidence study. *Oral Surgery Oral Medicine Oral Pathology* 50(3):287–291, 1980.

Corruccini, R.S.: An examination of the meaning of cranial discete traits for human skeletal biological studies. *American Journal of Physical Anthropology* 40:425–446, 1974.

Corruccini, R. S., Handler, J. S., and Jacobi, K. P.: Chronological distribution of enamel hypoplasias and weaning in a Caribbean slave population. *Human Biology* 57:699–711, 1985.

Cowell, M. J., and Cowell, H. R.: The incidence of spina bifida occulta in idiopathic scoliosis. *Clinical Orthopaedics and Related Research* 118:16–18, 1976.

Crain, J.B.: *Human paleopathology: A biliographic list.* Sacramento Anthropological Society, Paper 12, Sacramento, 1971.

Crane-Kramer, G. M. M.: *The paleopidemiological examination of treponemal infection and leprosy in Medieval populations from Northern Europe.* Ph.D. dissertation, Department of Archaeology. University of Calgary, 2000.

Crelin, E. S.: Development of the musculoskeletal system. *Clinical Symposia* 33(1):1–36, 1981.

Crosby, A. W.: The early history of syphilis: A reappraisal. *American Anthropology 71*:218–227, 1969.

Cucina, A., and Iscan, M. Y.: Assessment of enamel hypoplasia in a high status burial site. *American Journal of Human Biology 9*:213–222, 1997.

Cucina, A.: Brief communication: Diachronic investigation of linear enamel hypoplasia in prehistoric skeletal samples from Trentino, Italy. *American Journal of Physical Anthropology 119*:283–287, 2002.

Cummings, S. R., Kelsey, J. L., Nevitt, M. C., and O'Dowd, K. J.: Epidemiology of osteoporosis and osteoporotic fractures. *Epidemiologic Review 7*:178–208, 1985.

Cybulski, J. S.: Cribra orbitalia, a possible sign of anemia in early historic native populations of the British Columbia Coast. *American Journal of Physical Anthropology, 47*(1):31–39, 1977.

Cyron, B. M., and Hutton, W. C.: The fatigue strength of the lumbar neural arch in spondylolysis. *Journal of Bone and Joint Surgery, 60-B*(2):234–238, 1978.

Dahlberg, A.A.: The dentition of the American Indian. In Laughlin, W.S.: *The physical anthropology of the American Indian.* New York:Viking Fund, pp.138–76, 1951.

Dahlin, D.: *Bone tumors,* 2nd ed. Springfield: Thomas, 1967.

Dalinka, M. K., Dinnenberg, S., Greendyke, W. H., and Hopkins, R.: Roentgenographic graphic features of osseous coccidioidomycosis and differential diagnosis. *Journal of Bone and Joint Surgery 53-A*:1157–1164, 1971.

Dallman, P., Siimes, R., and Stekel, A.: Iron deficiency in infancy and childhood. *American Journal of Clinical Nutrition 33*:86–118, 1980.

D'Ambrosia, R.D.: Conservative management of metatarsal and heel pain in the adult foot. *Orthopedics 10*(1):137–142, 1987.

D'Ambrosia, R. D., and MacDonald, G. L.: Pitfalls in the diagnosis of Osgood-Schlatter disease. *Clinical Orthopedics 110*(Jul-Aug):206–209, 1975.

Daniels, E. G., and Nashel, D. J.: Periostitis: A manifestation of venous disease and skeletal hyperostosis. *Journal American Podiatric Association 73*(9):461–464, 1983.

Daveny, J. K., and Ross, M. D.: Cryptococcosis of bone. *Central African Journal of Medicine 15*:78–79, 1969.

David, D. J., Poswillo, D., and Simpson, D.: *The craniosynostoses: Causes, natural history and management.* Berlin, Springer-Verlag, 1982.

David, R., Oria, R. A., Kumar, R., Singleton, E. B., Lindell, M. M., Shirkhoda, A., and Madewell, J. E.: Radiologic features of eosinophilic granuloma of bone. *American Journal of Roentgenology 153*:1021–1026, 1989.

Davis, P. R.: The Thoraco-lumbar mortice joint. *Journal of Anatomy 89*:370–377, 1955.

Day, S. B.: Ossified subperiosteal hematoma. *Journal of American Medical Association 173*:986–990, 1960.

Dean, V.L.: *Effects of Cultural Deformation and Craniosynostosis on Cranial Venous Sinus and Middle Meningeal Vessel Pattern Expression.* Doctoral Dissertation, Indiana University, Bloomington, 1995.

Dee, P. M.: The preauricular sulcus. *Radiology 140*(2):354, 1981.

Deesomchok, U., and Tumrasvin, T.: Clinical comparison of patients with ankylosing spondylitis, Reiter's syndrome and psoriatic arthritis. *Journal Medical Association of Thailand 76*(2):61–70, 1993.

De Graaf, R. J., Matricali, B., and Hamburger, H. L.: Butterfly vertebra. *Clinics Neurology and Neurosurgery 84*(3):163–169, 1982.

Deguine, C., and Pulec, J.L.: Large osteoma of the external auditory canal. *Ear Nose and Throat Journal 80*(1):8, 2001.

___: Large nonobstructing exostoses of the external auditory canal. *Ear Nose and Throat Journal 80*(3):134, 2001.

Depalma, A. F.: Surgical anatomy of acromioclavicular and sternoclavicular joints. *Surgical Clinics of North America 43*(Dec):1541–1550, 1963.

De Smet, A .A., Ilahi, O. A., and Graf, B. K.: Untreated osteochondritis dissecans of the femoral condyles: Prediction of patient outcome using radiographic and MR findings.

*Skeletal Radiology 26:*463–467, 1997.

Derry, D. E.: Note on the accessory articular facets between the sacrum and ilium, and their significance. *Journal of Anatomy and Physiology 45:*202–210, 1911.

Desai, S. S., Patel, M. R., Michelli, L. J., Silver, J. W., and Lidge, R. T.: Osteochondritis Dissecans of the Patella. *Journal of Bone and Joint Surgery 69–B*(2):320–325, 1987.

Devoto, F. C.: Shovel-shaped incisors in Pre-Columbian Tastilian Indians. *Journal Dental Research 50*(1):168, 1971.

Devoto, F. C., Arias, N. H.: Shovel-shaped incisors in early Atacama Indians. *Journal Dental Research 46*(6):1478, 1967.

DiBartolomeo, J. R.: Exostoses of the external auditory canal. *Annals Otology Rhinology Laryngology Supplement 88*(6) Pt 2 Supplement 61:2–20, 1979.

Dickel, D. N., and Doran, G. H.: Severe neural tube defect syndrome from the Early Archaic of Florida. *American Journal of Physical Anthropology 80:*325–334, 1989.

Dickson, R. A.: Conservative treatment for idiopathic scoliosis. *Journal of Bone and Joint Surgery 67–B:*176–181, 1985.

Diekerhof, C. H., Reedt Dortland, R. W., Oner, F. C., and Verbout, A. J.: Severe erosion of lumbar vertebral body because of abdominal aortic false aneurysm: Report of two cases. *Spine 27*(16):E382–384, 2002.

Dieppe, P. A., Bacon, P.A., Bamji, A. N., and Watt, I.: *Atlas of Clinical Rheumatology.* Philadelphia: Lea & Febiger, 1986.

Di Maio, V. J. M.: *Gunshot wounds: Practical aspects of firearms, ballistics, and forensic techniques,* 2nd ed. Boca Raton: CRC Press, 1999.

Dingwall, E. J.: *Artificial cranial deformation.* London: John Bale, Sons and Danielson, Ltd., 1931.

Dixon, D. S.: Keyhole lesions in gunshot wounds of the skull and direction of fire. *Journal of Forensic Sciences 27:*555–556, 1982.

Dodo, Y.: Aural exostoses in the human skeletal remains excavated in Hokkaido. *Journal of the Anthropological Society, Nippon 80:*11–22, 1972.

Dolan, P., Earley M., and Adams, M. A.: Bending and compressive stresses acting on the lumbar spine during lifting activities. *Journal of Biomechanics 27:*1237–1248, 1994.

Donnelly, L. F., Bisset, G. S, Helms, C. A, and Squire, D. L: Chronic avulsive injuries of childhood. *Skeletal Radiology 28:*138–144, 1999.

Douglas, T. E., Jr.: Facial pain from elongated styloid process. *Archives of Otolaryngology 56:*635–638, 1952.

Dugdale, A. E., Lewis, A. N., and Canty, A. A.: The natural history of chronic otitis media. *New England Journal of Medicine 307:*1459–1460, 1982.

Dutour, O.: Enthesopathies as indicators of the activities of Neolithic Saharan populations. *American Journal of Physical Anthropology 71:*221–224, 1986.

Dutour, O., Palfi, G., Berato, J., and Brun, J-P. (Eds.) *The Origin of Syphilis in Europe Before or After 1493?* Paris: Editions Errance, 1994.

Dwight, T.: Account of two spines with cervical ribs, one of which cervical ribs, one of which has a vertebra suppressed, and absence of the anterior arch of the atlas. *Journal of Anatomy and Physiology 21:*539–550, 1887.

____: A bony supracondyloid process in the child. *American Journal of Anatomy 3:*221–228, 1904.

____: The clinical significance of variations of the wrist and ankle. *Journal of the American Medical Association 47:*252–255, 1906.

Dzurek, W.V., Etter, L. E., Keagy, R. M., and Miller M. W.: Enlarged parietal foramina. A collection of examples from radiologic practices in Pennsylvania. *Medical Radiography and Photography 32:*73–78, 1956.

Eagle, W. W.: Elongated styloid process: Further observations and a new syndrome. *Archives of Otolaryngology 47:* 630–640, 1948.

Echols, R. M., Selinger, D. D., Hallowell, C., Goodwin, J. S., Duncan, M. H., and Cushing, A. H.: Rhizopus osteomyelitis: A case report and review. *American Journal of Medicine*

66(1):141–145, 1979.

Edelson, J. G., Nathan, H., and Arensburg, B.: Diastematomyelia-the "double-barrelled" spine. *Journal of Bone and Joint Surgery 69–B*(2):188–189, 1987.

Edelson, J.G., Zuckerman, J. and Hershkovitz, I.: Os Acromiale: Anatomy and surgical implications. *Journal of Bone and Joint Surgery 75–B*:219–273,1993

Edwards, D. H., and Bentley, G.: Osteochondritis disecans patellae. *Journal of Bone and Joint Surgery, 59–B*(1):58–63, 1977.

Edwards, W. G.: Complications of suppurative otitis media. In Ludman, H. (Ed.): *Mawson's diseases of the ear,* 5th ed. Chicago: Year Book Medical Publishers. Inc., 1988.

Eggen, S., and Natvig, B.: Relationship between torus mandibularis and number of present teeth. *Scandinavian Journal of Dental Research 94*(3):233–240, 1986.

____: Concurrence of torus mandibularis and torus palatinus. *Scandinavian Journal of Dental Research 102*(1):60–63, 1994.

Eichenholtz, S. N.: *Charcot joints.* Springfield: Thomas, 1966.

Eisele, S. A., and Sammarco, G. J.: *Fatigure fractures of the foot and ankle in the athlete.* Journal of Bone and Joint Surgery 75(2):290–298, 1993.

Eisenstein, S.: Spondylolysis: A skeletal investigation of two populations. *Journal of Bone and Joint Surgery 60–B*:488–494, 1978.

Elerich, D. and Tyson, R. (Eds.) *Human paleopathology–Related subjects: An international bibliography.* San Diego, San Diego Museum of Man, 1997.

Ell, S. R.: Reconstructing the epidemiology of medieval leprosy: preliminary efforts with regard to Scandinavia. *Perspectives in Biology and Medicine 31*:496–506, 1988.

El Maghraoui, A., Tacache, F, El Khattabi, A., Bezza, A., Abouzahir, A., Ghafir, D., Ohayon, V., and Archane, M. I.: Abdominal aortic aneurysm with lumbar vertebral erosion in Behcet's disease revealed by low back pain: a case report and review of the literature. *Rheumatology (Oxford) 40*(4):472–473, 2001.

El-Najjar, M.Y. and Dawson, G.L.: The effects of artificuial cranial deformation on the incidence of wormian bones in the lambdoidal suture. *American Journal of Physical Anthropology 46*:155–160, 1977.

El-Najjar, M. Y., Lozoff, B., and Ryan, D. J.: The paleoepidemiology of porotic hyperostosis in the American Southwest: Radiological and ecological considerations. *American Journal of Roentgenology Radium Therapy Nuclear Medicine 125*(4):918–924, 1975.

El-Najjar, M., and Robertson, A. L Jr.: Spongy bones in prehistoric America. *Science 193*(4248):141–143, 1976.

El-Najjar, M. Y., Ryan, D. J., Turner, C. G., and Lozoff, B.: The etiology of porotic hyperostosis among the prehistoric and historic Anasazi Indians of southwestern United States. *American Journal of Physical Anthropology 44*:477–488, 1976.

Epstein, B. S.: The concurrence of parietal thinness with postmenopausal, senile or idiopathic osteoporosis. *Radiology 60*:29–35, 1953.

Epstein, H. C.: *The spine: A radiological text and atlas* (4th ed.). Philadelphia, Lea and Febiger, 1976.

Erken, E., Ozer, H. T., Gulek, B., and Durgun, B.: The association between cervical rib and sacralization. *Spine 27*(15):1659–1664, 2002.

Eshed, V., Latimer, B., Greenwald, C. M., Jellema, L. M., Rothschild, B. M., Wish-Baratz, S., and Hershkovitz, I.: Button osteoma: Its etiology and pathophysiology. *American Journal of Physical Anthropology 118*(3):217–230, 2002.

Fairgrieve, S. I., and Molto, J. E.: Cribra orbitalia in two temporally disjunct population samples from the Dakhleh Oasis, Egypt. *American Journal of Physical Anthropology 111*(3):319–331, 2000.

Farfan, H. F., Osteria, V., and Lamy C.: The mechanical etiology of spondylolysis and spondylolisthesis. *Clinical Orthopedics 117*:40–55, 1976.

Farkas, A.: Physiological scoliosis. *Journal of Bone and Joint Surgery 23*(3):607, 1941.

Fehlandt, A. F., and Micheli, L. J.: Lumbar facet stress fracture in a ballet dancer. *Spine 18*(16):2537–2539, 1993.

Fein, J.M. and Brinker, R.A.: Evolution and significance of giant parietal foramina. *Journal of Neurosurgery 37:*487–492, 1972.

Feldman, V. B.: Eagle's syndrome: A case of symptomatic calcification of the stylohyoid ligaments. *Journal Canadian Chiropractic Association 47*(1):21–27, 2003.

Felson, D. T., Naimark, A. Anderson, J. Kazis, L., Castelli, W. and Meenan, R. F.: The prevalence of knee osteoarthritis in the elderly: The Framingham Osteoarthritis Study. *Arthritis and Rheumatism 30*(8):914–918, 1987.

Fenton, J. E., Turner, J., and Fagan, P. A.: A histopathologic review of temporal bone exostoses and osteomata. *Laryngoscope 106*(5 Pt 1):624–628, 1996.

Ferembach, D.: Frequency of spina bifida occulta in prehistoric human skeletons. *Nature 199:*100–101, 1963.

Ferrario, V. F., Sigurta, D., Daddona, A., Dalloca, L., Miani, A., Tafuro, F., and Sforza, C.: Calcification of the stylohyoid ligament: Incidence and morphoquantitative evaluations. *Oral Surgery Oral Medicine Oral Pathology 69*(4):524–529, 1990.

Fiese, M. J.: *Coccidioidomycosis.* Springfield: Thomas, 1958.

Filipo, R., Rabiani, M., and Barbara, M.: External ear canal exostosis: A physiopathological lesion in aquatic sports. *Journal of Sports Medicine Physical Fitness 22*(3):329–336, 1982.

Fink, I. J., Pastakia, B., and Barranger, J. A.: Enlarged phalangeal nutrient foramina in Gaucher disease and B-thalassemia major. *American Journal of Roentgenology 143:*647–649, 1984.

Finnegan, M., and Faust, M. A.: *Bibliography of human and nonhuman non-metric variation.* Research Reports N. 14, Department of Anthropology, University of Massachusetts. Amherst, University of Massachusetts, 1974.

Finnegan, M., and Marcsik, A. M.: Anatomy or pathology: The Stafine Defect as seen in archaeological material and modern clinical practice. *Journal of Human Evolution 9:*19–31, 1980.

Fiorato, V., Boylston, A., and Knusel, C.: *Blood red roses: The archaeology of a mass grave from the Battler of Towton, AD 1461.* Oxford:Oxbow Books, 2001.

Fishbien, M. (Ed.): *Birth defects.* Philadelphia:J.B. Lippencott, 1963.

Fitoz, S., Atasoy, C., Yagmurlu, A., and Akyar, S.: Psoas abscess secondary to tuberculous lymphadenopathy: case report. *Abdominal Imaging 26*(3):323–324, 2001.

Fitton, J. S.: A tooth ablation custom occuring in the Maldives. *British Dental Journal 175*(8):299–300, 1993.

Fiumara, N. J., and Lessell, S.: Manifestation of late congenital syphilis: An analysis of 271 patients. *Archives of Dermatology 102:*78–83, 1970.

____: The stigmata of late congenital syphilis: and analysis of 100 patients. *Sexually Transmitted Diseases 10*(3):126–129, 1983.

Fletcher, H. A, Donoghue, H. D, Holton, J., Pap, I., and Spigelman, M.: Widespread Occurrence of Mycobacterium tuberculosis DNA Form 18th - 19th Century Hungarians. *American Journal of Physical Anthropology 120:*144–152, 2003.

Floman, Y., Bloom, R. A., and Robin, G. C.: Spondylolysis in the upper lumbar spine: A study of 32 patients. *Journal of Bone and Joint Surgery 69-B*(4):582, 1987.

Foreman, S. M.: Fractures and Dislocations of the Cervical Spine. In Foreman, S. M. and Croft, A. C. (Eds.): *Whiplash injuries: The cervical acceleration / deceleration syndrome.* Lippincott Williams & Wilkins, Philadelphia, pp. 271–295, 2001.

Fowler, F. D., Driebe, W. T., and Copeland, M. M.: Suprcondyloid process. *The American Surgeon 25*(6):266–272, 1959.

Fox, R. J., Walji, A. H., Mielke, B., Petruk, K.C., and Aronyke, K. E.: Anatomic details of intradural channels in the parasagittal dura: A possible pathway for flow of cerebrospinal fluid. *Neurosurgery 39*(1):84–90, 1996.

Franco-Paredes, C., and Blumberg, H. M.: Psoas muscle abscess caused by Mycobacterium tuberculosis and Staphylococcus aureus: Case report and review. *American Journal Medical Science 321*(6):415–417, 2001.

Francois, R. J.: Microradiographic study of the intervertebral bridges in ankylosing spondylitis and in the normal sacrum. *Annals of Rheumatic Diseases 24*:481–489, 1965.

Fredrickson, B. E., Baker, D., McHolick, W. J., Yuan, H. A., and Lubicky, J. P.: The natural history of spondylolysis and spondylolisthesis. *Journal of Bone and Joint Surgery 66–A*:699–707, 1984.

Friedman, R. J., and Micheli, L. J.: Acquired spondylolisthesis following scoliosis surgery. A case report. *Clinical Orthopaedics 190*:132–134, 1984.

Frymoyer, J. W.: Back pain and sciatica. *New England Journal of Medicine 318*(5):291–300, 1988.

Galloway, A. (Ed.): *Broken bones: Anthropological analysis of blunt force trauma.* Springfield: C. C Thomas, 1999.

Ganguili, P. K.: *Radiology of bone and joint tuberculosis.* New York: Asia Publishing House, 1963.

Garber, E. A., and Silver, S.: Pedal manifestations of DISH. *Foot and Ankle 3*(1):12–16, 1982.

Garcia-Lechuz, J. M., Julve, R., Alcala, L., Ruiz-Serrano, M. J., and Munoz, P.: Tuberculous spondylodiskitis (Pott's disease): Experience in a general hospital [Article in Spanish, English abstract]. *Enferm Infec Microbiol Clin 20*(1):5–9, 2002.

Gardner, D. L.: *Pathology of the connective tissue diseases.* London: Edward Arnold, 1965.

Garth, W. P., and Van Patten, P. K.: Fractures of the lumbar lamina with epidural hematoma simulating herniation of a disc: A case report. *Journal of Bone and Joint Surgery 71–A*:771–772, 1989.

Genant, H. K.: Osteoporosis and bone mineral assessment. In McCarty, D. J. (Ed.): *Arthritis and allied conditions: A textbook of rheumatology,* 11th ed. Philadelphia, Lea and Febiger, 1989.

Genner, B. A.: Fracture of the supracondyloid process. *Journal of Bone and Joint Surgery 41A*:1333–1335, 1959.

Gensburg, R. S., Kawashima, A., and Sandler, C. M.: Scintigraphic demonstration of lower extremity periostitis secondary to venous insufficiency. *Journal of Nuclear Medicine 29(7)*:1279–1282, 1988.

George, G. R.: Bilateral bipartite patellae. *British Journal of Surgery 22*:555, 1935.

Gerber, N., Ambrosini, G. C., Boni, A., Fehr, K., and Wagenhauser, F. J.: Ankylosing spondylitis (Bechterew) and tissue antigen HLA–B27. II. HLA–b27 negativity in classical clinical ankylosing spondylitis no independent nosological entity (in German). *Zeitschrift fur Rheumatologie 36*(7):124–129, 1977.

Gerbino, P. G., and d'Hemecourt, P. A.: Does football cause an increase in degenerative disease of the lumbar spine? *Current Sports Medicine Reports 1*(1):47–51, 2002.

Gerbino, P. G., and Micheli, L. J.: back injuries in the young athlete. *Clinics in Sports Medicine 14*:571–590, 1995.

Gerschon-Cohen, J., Schraer, H., Blumberg, N.: Hyperostosis frontalis interna among the aged. *American Journal of Roentgenology 73*:396–397, 1955.

Gerscovich, E, O., Greenspan, A., and Szabo, R. M.: Benign clavicular lesions that may mimic malignancy. *Skeletal Radiology 20*:173–180, 1991.

Gerszten, P.C.: An investigation into the practice of cranial deformation among the pre-Columbian peoples of northern Chile. *International Journal of Osteoarchaeology 3*:87–98, 1993.

Gerszten, P. C., Gerszten, E., and Allison, M. J.: Diseases of the skull in pre-Columbian South American mummies. *Neurosurgery 42*(5):1145–1151, 1998.

Gibson, M. J., Szypryt, E. P., Buckley, J. H., Worthington, B. S., and Mulholland, R. C.: Magnetic resonance imaging of adolescent disc herniation. *Journal of Bone and Joint Surgery 69–B*(5):699–703, 1987.

Giles, R. G.: A congenital anomaly of the patella. *Texas State Journal of Medicine 23*:731–732, 1928.

Gindhart, P. S.: The frequency of transverse lines in the tibia in relation to childhood illness. *American Journal of Physical Anthropology 31*:17–22, 1969.

Giroux, J. C., and Leclerq, T. A.: Lumbar disc excision in the second decade. *Spine 7*:168–170, 1982.

Gladstone, R. J., and Wakeley, C. P. G.: Cervical ribs and rudimentary first thoracic ribs considered from clinical and etiological standpoints. *Journal of Anatomy 66*:334–370, 1931–32.

Glen-Haduch, E., Szostek, K., and Glab, H.: Cribra orbitalia and trace element content in human teeth from Neolithic and Early Bronze Age graves in southern Poland. *American Journal of Physical Anthropology, 103*(2):201–207, 1997.

Goff, C. W.: Syphilis. In Brothwell, D., and Sandison, A. (Editors). *Diseases in Antiquity.* Springfield, IL: Charles C Thomas, pp. 279–293, 1967.

Goldsand, G.: Actinomycosis. In Hoeprich, P. D. and Jordan, M. C. (Eds.): *Infectious disease: A modern treatise of infectious processes,* 4th ed. Philadelphia, Lippincott, pp. 666–684, 1989.

Goldsmith, W. M.: The Catlin Mark: the inheritance of an unusual opening in the parietal bones. *Journal of Heredity 13:*69–71, 1922.

Goodman, A. H., Allen, L. H., Hernandez, G. P., Arriola, A., Chavez, A., and Pelto, G. H.: Prevalence and age at development of enamel hypoplasias in Mexican children. *American Journal of Physical Anthropology 72:*7–19, 1987.

Goodman, A. H., and Armelagos, G. J.: Factors affecting the distribution of enamel hypoplasia in human permanent dentition. *American Journal of Physical Anthropology 68:*479–493, 1985.

Goodman, A. H., and Armelagos, G. J.: Childhood stress and decreased longevity in a prehistoric population. *American Anthropologist 90*(4):936–944, 1988.

Goodman, A. H., Armelagos, G. J., and Rose, J. C.: Enamel hypoplasia as indicator of stress in three prehistoric populations from Illinois. *Human Biology 52:*515–528, 1980.

Goodman, A. H. and Clark, G. A.: Harris lines as indicators of stress in prehistoric Illinois populations. In Martin, D. L. and Bumsted, M. P. (Eds.) *Biocultural Adaptation: Comprehensive Approaches to Skeletal Analysis.* Research Reports No. 20. Amherst, University of Massachusetts, pp. 35–46, 1981.

Goodman, A. H., and Rose, J. C.: Assessment of systemic physiological perturbations from dental enamel hypoplasias and associated histological structures. *Yearbook of Physical Anthropology 33:*59–110, 1990.

Goodman, A. H., and Rose, J. C.: Dental enamel hypoplasias as indicators of nutritional status. In Kelley, M. A., and Larsen, C. S. (Editors). *Advances in Dental Anthropology.* New York: Wiley Liss, pp. 279–293, 1991.

Goodman, A. H., and Song, R. J.: Sources of variation in estimated ages of formation of linear enamel hypoplasias. In Hoppa, R. D., and Fitzgerald, C. M (Editors). *Human growth in the past.* Cambridge: Cambridge University Press, pp. 210–240, 1999.

Goodman, R. M. and Gorlin, R. J.: *Atlas of the face in genetic disorders.* St. Louis:C.V. Mosby. 1977.

Gorab, G. N., Brahney, C., and Aria, A. A.: Unusual presentation of a Stafne bone cyst. *Oral Surgery Oral Medicine Oral Pathology 61*(3):213–215, 1986.

Gorlin, R. J., Pindborg, J. J., and Cohen, M. M.: *Syndromes of the head and neck.* New York:McGraw-Hill, 1976.

Gorsky, M., Bukai, A., and Shohat, M.: Genetic influence on the prevalence of torus palatinus. *American Journal Medical Genetics 75*(2):138–140, 1998.

Gould, A. R., Farman, A. G., and Corbitt, D.: Mutilations of the dentition in Africa: a review with personal observations. *Quintessence International 15*(1):89–94, 1984.

Graham, M. D.: Osteomas and exostoses of the external auditory canal. A clinical histopathologic and scanning electron microscopic study. *Annals Otology Rhinology Laryngology 88*(4 Pt 1):566–572, 1979.

Grainger, R. G., Allison, D. J., Adam, A., and Dixon, A. K. (Eds.): *Diagnostic radiology: A textbook of medical imaging,* 4th ed. Churchill Livingstone, London, 2001.

Grandmaison, G. L. de la, Brion, F., and Durigon, M.: Frequency of bone lesions: An inadequate criterion for gunshot wound diagnosis in skeletal remains. *Journal of Forensic Sciences 46*(3):593–595, 2000.

Grant, J. C. B.: *A method of anatomy: Descriptive and deductive,* 4th ed. Baltimore: Williams and Wilkins, 1948.

____: *Grant's atlas of anatomy,* 6th ed. Baltimore, Williams and Wilkins, 1972.

Grauer, A. and Roberts, C.: Paleoepidemiology, healing and possible treatment of trauma in the medieval cemetery population of St. Helen-on-the-Walls, York England. *American Journal of Physical Anthropology 100:*531–544, 1996.

Graves, W. W.: Observations on age changes in the scapula. *American Journal of Physical Anthropology 5:*21–33, 1922.

Gray, H.: *Anatomy of the human body,* 25th Ed. Philadelphia: Lea and Febiger, 1948.

Green, W. T. Jr.: Painful bipartite patellae: A report of three cases. *Clinical Orthopaedics and Related Research 110:*197–200, 1975.

Greenfield, G. B.: *Radiology of bone diseases,* 2nd ed. Philadelphia: Lippincott, 1975.

Greer, A. E.: *Disseminating fungal diseases of the lung.* Springfield: Thomas, 1962.

Gregg, J. B., and Bass, W. M.: Exostoses in the external auditory canals. *Annals of Otology Rhinology and Laryngology 79:*834–839, 1970.

Gregg, J. B., and Gregg, P. S.: *Dry bones: Dakota Territory reflected.* Sioux Falls, Sioux Falls Printing, 1987.

Griffiths, H. J.: *Basic bone pathology.* Norwalk: Appleton-Century-Crofts, 1981.

Grolleau-Raoux, J-L., Crubezy, D. R., Brugne, J-F, and Saunders, S. R.: Harris Lines: A Study of Age-Associated Bias in Counting and Interpretation. *American Journal of Physical Anthropology 103:*209–217, 1997.

Gruber, W.: Ueber einen neuen sekundarrn Tarsalknochen-Calcaneus secundariusmit Bemerkungen uber den Tarsus uberhaupt. *Memoirs de l'Academic des Science de St. Petersbourg,* T. 17, No. 6, 1871.

Grupe, G.: Metastasizing carcinoma in a medieval skeleton: Differential diagnosis and etiolgy. *American Journal of Physical Anthropology 75:*369–374, 1988.

Guidotti, A.: Frequencies of cribra orbitalia in central Italy (19th century) under special consideration of their degrees of expression. *Anthropology Anz, 42*(1): 11–16, 1984.

Gulekon, I. N., and Turgut, H. B.: The preauricular sulcus: Its radiologic evidence and prevalence. *Kaiboguku Zasshi, 76*(6):533–535, 2001.

____: The external occipital protuberance: Can it be used as a criterion in the determination of sex? *Journal of Forensic Sciences 48*(3):513–516, 2003.

Guyuron, B., Uzzo, C. D., and Scull, H.: A practical classification of septonasal deviation and an effective guide to septal surgery. *Plastic Reconstructive Surgery 104*(7):2202–2209, 1999.

Hackett, C. J.: On the origin of the human treponematoses. *Bull WHO 29:*7–41, 1963.

Hackett, C. J.: The human treponematoses. In Brothwell, D. R. and Sandison, A. T. (Eds.): *Diseases in antiquity: A survey of the diseases injuries and surgery of early populations,* Springfield: Thomas, pp. 152–169, 1967.

Hackett, C. J.: An introduction to diagnostic criteria of syphilis, treponarid and yaws (treponematoses) in dry bones, and some implications. *Virchows Archives Pathology Anatomy Histology, 368*(3):229–241, 1975.

____: Diagnostic criteria of syphilis, yaws and treponarid (treponematoses) and some other diseases in dry bone. *Sitzunosbericht der heidelberger Akademic der Wissenschaft 4,* 1976.

____: Development of caries sicca in a dry calvaria. *Virchows Archiv. A, Pathological Anatomy and Histology 391*(1):53–79, 1981.

Hackett, C. J.: Problems in the palaeopathology of the human treponematoses. In Hart, G. D. (editor). *Disease in Ancient Man.* Toronto: Clarke Irwin, pp. 106–128, 1983.

Hagberg, M., and Wegman, D. H.: Prevalence rates and odds ratios of shoulder-neck diseases in different occupational groups. *British Journal of Indian Medicine 44:*602–610, 1987.

Hahn, P. Y., Strobel, J. J., and Hahn, F. J.: Verification of lumbosacral segments on MR images: identification of transitional vertebrae. *Radiology 182*(2):615–616, 1992.

Haibach, H., Farrell, C., and Gaines, R. W.: Osteoid osteoma of the spine: Surgically correctable cause of painful scoliosis. *Canadian Medical Association Journal 135*(8):895–899, 1986.

Haidar, A., and Kalamchi, S.: Painful dysphagia due to fracture of the styloid process. *Oral Surgery Oral Medicine Oral Pathology 49*(1):5–6, 1980.

Hallock, H. and Jones, J. B.: Tuberculosis of the spine. *Journal of Bone and Joint Surgery 36:*219–240, 1954.

Halpern, A A., and Hewitt, O.: Painful medial bipartite patellae: A case report. *Clinical Orthopaedics and Related Research 134*:180–181, 1978.

Hamilton, W. C. (Ed.).: *Traumatic Disorders of the Ankle.* New York, Springer-Verlag, 1984.

Handler, J. S., Corruccini, R. S., and Mutaw, R. J.: Tooth mutilation in the Caribbean: Evidence from a slave burial population in Barbados. *Journal of Human Evolution 11*:297–313, 1982.

Hanihara, T., and Ishida, H.: Frequency variations of discrete cranial traits in major human populations. III. Hyperostotic variations. *Journal of Anatomy 199*(Pt3):251–272, 2001 (a).

Hanihara, T., and Ishida, H.: Os incae: Variation in frequency in major human population groups. *Journal Anatomy 198*(Pt 2):137–152, 2001 (b).

Hansen, G. L.: Pers. Comm., Smithsonian Institution, Washington, D. C., 1989.

Hansson, L. G.: Development of a lingual mandibular bone cavity in an 11-year-old boy. *Oral Surgery Oral Medicine Oral Pathology 49*(4):376–378, 1980.

Harbert, J., and Desai, R.: Small calvarial bone scan foci–Normal variations. *Journal Nuclear Medicine 26*(10):1144–1148, 1985.

Harris, H. A.: The growth of the long bones in childhood: With special reference to certain bony striations of the metaphysis and the role of vitamins. *Archives of Internal Medicine 38*:622–640, 1926.

____: Lines of arrested growth in long bones in childhood. Correlation of histological and radiographic appearances in clinical and experimental conditions. *British Journal of Radiology 4*:561–588, 1931.

____: *Bone growth in health and disease: The biological principles underlying the clinical, radiological and histological diagnosis of perversions of growth and disease in the skeleton.* London: Oxford University Press, 1933.

Harris, R. I., and Wiley, J. J.: Acquired spondylolysis as a sequel to spine fusion. *Journal of Bone and Joint Surgery 45-A*:1159–1170, 1963.

Harverson, G., and Warren A. G.: Tarsal bone disintegration in leprosy. *Clinical Radiology 30*(3):317–322, 1979.

Harvey, W., and Noble, H. W.: Defects on the lingual surface of the mandible near the angle. *British Journal Oral Surgery 6*:75–83, 1968.

Hatfield, K. D.: The preauricular sulcus. *Australas Radiology 15*(2):168–169, 1971.

Hauser, G., and DeStefano, G. F (Eds.): Introduction. In *Epigenetic variants of the human skull.* Stuttgart: Schweizerbart, 1989.

Haverstock, B. D.: Anterior ankle abutment. *Clinics Podiatric Medicine Surgery 18*(3):457–465, 2001.

Hawkey, D. E.: Disability, compassion and the skeletal record: Using musculoskeletal stress markers (MSM) to construct and osteobiography from early New Mexico. *International Journal of Osteoarchaeology 8*(5):326–340, 1998.

Hawkins, R. B.: Arthroscopic treatment of sports-related anterior osteophytes in the ankle. *Foot and Ankle 9*(2):87–90, 1988.

Hellman, M. H.: Charcot joint disease (Charcot joints). In McCarty, D. J. (Ed.): *Arthritis and allied conditions: A textbook of rheumatology,* 11th ed. Philadelphia: Lea and Febiger, 1989.

Hengen, O. P.: Cribra orbitalia: pathogenesis and probable etiology. *HOMO 22*:57–75, 1971.

Henrard, J. C., and Bennett, P. H.: Etude epidemiologique de l'hyperostose vertebrale. Enquete dans une population adulte d'Indiens d'Amerique. *Revue du Thumatisme et des Maladies Osteoarticulaires 40*:581–591, 1973.

Hergan K., Oser, W., and Moriggl B.: Acetabular ossicles: Normal variant or disease entity? *European Radiology, 10*(4): 624–628, 2000.

Hertslet, L. E., and Keith, A.: Comparison of anomalous parts of two subjects, one with a cervical rib, the other with a rudimentary first rib. *Journal of Anatomy and Physiology 30*:562–567, 1896.

Hertzler, A. E.: Surgical pathology of the diseases of bone. *Hertzler's Monographs on Surgical Pathology.* Chicago: Lakeside, 1931.

Herzog, S., and Fiese, R.: Persistent foramen of Huschke: possible risk factor for otologic complications after arthroscopy of the temporomandibular joint. *Oral Surgery Oral Medicine Oral Pathology 68*(3):267–270, 1989.

Hess, L.: The metopic suture and the metopic syndrome. *Human Biology 17:*107–136, 1945.

Hilel, N.: The para-articular processes of the thoracic vertebrae. *Journal of Anatomy 133:*605, 1959.

____: Osteophytes of the vertebral column: An anatomical study of their development according to age, race, and sex with considerations as to their etiology and significance. *Journal of Bone and Joint Surgery 44-A:*243, 1962.

Hill, C. L., Chaisson, C. E., Skinner, K., Kazis, L. Gale, M. E., and Felson, D. T.: Periarticular lesions detected on magnetic resonance imaging: Prevalence in knees with and without symptoms. *Arthritis and Rheumatism 48*(10):2836–2844, 2003.

Hill, M. C.: Porotic hyperostosis: A question of correlations verses causality. *American Journal of Physical Anthropology 66:*182, 1985.

Hillson, S. W., and Bond, S.: Relationship of enamel hypoplasia to the pattern of tooth crown growth: A discussion. *American Journal of Physical Anthropology 104:*89–103, 1997.

Hillson, S., Grigson, C., and Bond, S.: Dental defects of congenital syphilis. *American Journal of Physical Anthropology 1071:*25–40, 1998.

Hinkes, M. J.: Shovel shaped incisors in human identification. In Gill, G. W. and Rhine, S. (Eds.) *Skeletal attribution of race.* Anthropology Papers No. 4, Maxwell Museum of Anthropology 21–26, 1990.

Hirata, K.: A contribution to the paleopathology of of cribra orbitalia in Japanese. 1. Cribra orbitalia in Edo Japanese. *St. Marianna Medical Journal 16:*6–24, 1988.

Hitchcock, H. H.: Spondylolisthesis: Observations on its development, progression, and genesis. *Journal of Bone and Joint Surgery 22-B:*1, 1940.

Hoeprich, P. D.: Coccidioidomycosis. In Hoeprich, P. D. and M. C. Jordan (Eds.): *Infectious diseases: A modern treatise of infectious processes,* 4th ed. Philadelphia: Lippincott, pp. 489–502, 1989 (a).

____: Nonspecific treponematoses. Inn Hoeprich, P. D. and M. C. Jordan (Eds.): *Infectious diseases: A modern treatise of infectious process,* 4 th Ed. Philadelphia: Lippincott, pp. 1021–1034, 1989 (b).

Hodge, J. C.: Anterior process fracture of calcaneus secundarius: A case report. *Journal of Emergency Medicine 17*(2):305–309, 1999.

Hodgson, A. R., Wong, W., and Yau, A.: *X-Ray appearance of tuberculosis of the spine.* Springfield: C. C Thomas, 1969.

Hoffman, J. M.: Enlarged parietal foramina–their morphological variation and use in assessing prehistoric biological relationships. In Hoffman. J. M. and Brunker, L. (Eds.): *Studies in California Paleopathology,* Contributions of the University of California Archaeological Research Facility, No. 30. Berkeley, University of California, 1976.

Holcomb, R. C.: *Who gave the world syphilis? The Haitian myth.* New York, Froben, 1930.

____: The antiquity of congenital syphilis. *Medical Life 42:*275–325, 1935.

Hollender, L.: Enlarged parietal foramina. *Oral Surgery 23:*447–453, 1967.

Holt, C. A.: A re-examination of parturition scars on the human female pelvis. *American Journal of Physical Anthropology 49*(1):91–94, 1978.

Hooton, E. A.: *The Indians of Pecos Pueblo, A study of their skeletal remains.* Papers of the Southwestern Expedition No. 4. New Haven: Yale University Press, 1930.

Hoppenfeld, S.: Back pain. *Pediatric Clinics of North America 24*(4):881–887, 1977.

Horal, J., Nachemson, A., and Scheller, S.: Clinical and radiological long-term follow-up of vertebral fractures in childern. *Acta Orthopaedica Scandinavica 43:*491–503, 1972.

Hough, A. J., and Sokoloff, L.: Pathology of osteoarthritis. In McCarty, D. J. (Ed.): *Arthritis and allied conditions: A textbook of rheumatology,* 11th ed. Philadelphia, Lea and Febiger, 1989.

Houghton, P.: The relationship of the pre-auricular groove of the ilium to pregnancy. *American Journal of Physical Anthropology 41:*381–389, 1974.

____: The bony imprint of pregnancy. *Bulletin of the New York Academy of Medicine 51:*655–661, 1975.

Hrdlicka, A.: Artifical deformations of the human skull with special reference to America. In Lehmann-Nitsche, R.: *Actas del XVII Congreso Internacionale de Americanistas.* International Society of the Americanists, pp. 147–149, 1912.

____: *Anthropological work in Peru in 1913. With notes on the pathology of the ancient Peruvians.* Smithsonian Miscellaneous Collections 61, No. 18:vi–69, 1914.

____: Shovel-shaped teeth. *American Journal of Physical Anthropology 3*:429–465, 1920.

____: Incidence of the supercondyloid process in Whites and other races. *American Journal of Physical Anthropology 6*:405–412, 1923.

____: The principal dimensions, absolute and relative, of the humerus in the white race. *American Journal of Physical Anthropology 16*:431–450, 1932.

____: Ear Exostoses. *Smithsonian Miscellaneous Collections. 93*:1–100, 1935.

Hsu, J. W., Tsai, P. L., Hsiao, T. H., Chang, H. P., Lin, L. M., Liu, K. M., Yu, H. S., and Ferguson, D.: The effect of shovel trait on Carabelli's trait in Taiwan Chinese and Aboriginal populations. *Journal of Forensic Science 42*(5):802–806, 1997.

Hudson, E. H.: A footnote on yaws and syphilis; same or different? *Navy Medical Bulletin 38*:172–176, 1940.

____: *Non-venereal Syphilis: A sociological and medical study of bejel.* Edinburgh, E. & S. Livingstone, 1958.

Hughston, J. C., Hergenroeder, P. T., and Courtenay, B. G.: Osteochondritis dissecans of the femoral condyles. *Journal of Bone and Joint Surgery 66–A*(9):1340–1348, 1984.

Hutchinson, D. L., Denise, C. B., Daniel H. J., and Kalmus, G. W.: A reevaluation of the cold water etiology of external auditory exostoses. *American Journal of Physical Anthropology 103*(3):417–422, 1997.

Hutchinson, D. L., and Larsen, C. S.: Determination of stress episode duration from linear enamel hypoplasias: A case study from St. Catherine's Island, Georgia. *Human Biology 60*:93–110, 1988.

Hutter, R., Worcester, J., Francis, K., Foote, F., and Stewart, F.: Benign and malignant giant cell tumors of bone. *Cancer 15*:653–690, 1962.

Ihle, C. L., and Cochran, R. M.: Fracture of the fused os trigonum. *American Journal Sports Medicine 10*(1):47–50, 1982.

Iscan, M. Y. and Kennedy, K. A. R. (Eds.): *Reconstruction of life from the skeleton.* New York: Alan R. Liss, Inc. 1989.

Jackson, R.: (1988), Scoliosis in juvenile and adolescent children. *Health Visitor 61*(3):76–7.

Jaffe, H. L.: *Tumors and tumorous conditions of bones and joints.* Philadelphia: Lea and Febiger, 1958.

____: Ischemic necrosis of bone. *Medical Radiography and Photography 86*:58–86, 1969.

____: *Metabolic, degenerative and inflammatory diseases of bone and joints.* Philadelphia: Lea and Febiger, 1975.

Jager, H. J., Gordon-Harris, L., Mehring, U. M., Goetz, G. F, Mathias, K. D.: Degenerative change in the cervical spine and load-carrying on the head. *Skeletal Radiology 26*:475–481, 1997.

Jahss, M. H.: *Disorders of the foot and ankle: Medical and surgical management.* Philadelphia: W. B. Saunders Company, 1991.

Jainkittivong, A., and Langlais, R. P.: Buccal and palatal exostoses: prevalence and concurrence with tori. *Oral Surgery Oral Medicine Oral Pathology Oral Radiology Endodontics 90*(1):48–53, 2000.

James, R. and Nasmyth-Jones, R.: The occurrence of cervical fractures in victims of judicial hanging. *Forensic Science International 54*:81–91, 1992.

Jarcho, S. (Ed.): *Human paleopathology.* New Haven:Yale University Press, 1966.

Jayson, M. I. V., Dixon, A. J. (Eds).: *The lumbar spine and back pain* (4th ed.). New York: Churchill Livingston, 1992.

Jeffery, A. K.: Osteophytes and the osteoarthritic femoral head. *Journal of Bone and Joint Surgery 57–B*(3):314–324, 1975.

Jeyasingh, P., Gupta, C. D., Arora, A. K., and Ajmani, M. L.: Incidence of squatting facets on the talus of Indians (Agra region). *Anthropologischer Anzeiger 37*(2):117–122, 1979.

Jit, I., and Kaur, H.: Rhomboid fossa in the clavicles of North Indians. *American Journal of Physical Anthropology 70*:97–103, 1986.

Johansson, L. U.: Bone and related materials. In Hodge, W. M. (Ed.): *In situ archaeological conservation*. Mexico: Instituto Nacional de Antropologia e Historia, 1987.

Johnston, M. C., and Millicovsky, G.: Normal and abnormal development of the lip and palate. *Clinics Plastic Surgery 12*(4):521–532, 1985.

Jones, H. C., and Hedrick, D. W.: Patellar anomalies, roentgenologic and clinical consideration. *Radiology 38*:30–34, 1942.

Jones, R. R., and Martin, D. D.: Blastomycosis of bone. *Surgery 10*:931, 1941.

Julkunen, H., Heinonen, O. P., Knekt, P., and Maatela, J.: The epidemiology of hyperostosis of the spine together with its symptoms and related mortality in general population. *Scandinavian Journal of Rheumatology 40*:581, 1973.

Julkunen, H., Heinonen, O. P., and Pyorala, K.: Hyperostosis of the spine in an adult population: Its relation to hyperglycaemia and obestity. *Annals of Rheumatic Disease 30*:605–612, 1971.

Junge, A., El-Sheik, M., Celik, I., and Gotzen, L.: Pathomorphology, diagnosis and treatment of "hangman's fractures" (article in German). *Unfallchirung 105*(9):775–782, 2002.

Jurmain, R.: Paleoepidemiolgical patterns of trauma in a prehistoric population from central California. *American Journal of Physical Anthropology 115*:13–23, 2001.

Kaar, S. G., Cooperman, D. R., Blakemore, L, C., Thompson, G. H., Petersilge, C. A., Elder, J. S., and Heiple, K. G.: Association of bladder exstrophy with congenital pathology of the hip and lumbosacral spine: A long-term follow-up study of 13 patients. *Journal of Pediatric Orthopaedics 22*(1):62–66, 2002.

Kahn M. A.: Ankylosing spondylitis. In A. Calin (Ed.), *Spondylorthropathies*. Orlando: Grune & Stratton, 1984.

Karasick, D., and Schweitzer, M. E.: The os trigonum syndrome: Imaging features. *American Journal of Roentgenology 166*(1):125–129, 1996.

Karpinski, M. R. K., Newton, G., and Henry, A. P. J.: The results and morbidity of varus osteotomy for Perthes disease. In Burwell, R. G. and Harrison, M. H. M. (Eds.): *Clinical Orthopaedics and Related Research 209*:30–40, 1986.

Kate, B. R.: The incidence and cause of cervical fossa in Indian femora. *Journal of the Anatomical Society of India 12*(2):69, 1963.

Kate, B.R. and Robert, S.L.: Some observations on the upper end of the tibia in squatters. *Journal of Anatomy 99*:137–141, 1965.

Kaufmann, G. A., Sundarum, M., and Mcdonald, D. J.: Magnetic resonance imaging in symptomatic Paget's disease. *Skeletal Radiology 20*:413–418, 1991.

Kawashima, T., and Uhthoff, H. K.: Prenatal Development Around the Sustentaculum Tali and Its Relation to Talocalcaneal Coalitions. *Journal of Pediatric Orthopaedics 10*:238–243, 1990.

Kay, D. J., Har-El, G., and Lucente, F. E.: A complete stylohyoid bone with a stylohyoid joint. *American Journal of Otolaryngology 22*(5):358–361, 2001.

Kayalioglu, G., Oyar, O., and Govsa, F.: Nasal cavity and paranasal sinus bony variations: a computed tomographic study. *Rhinology 38*(3):108–113, 2000.

Keats, T. E.: *An atlas of normal roentgen variants that may simulate disease* (3rd ed.). Chicago: Year Book Medical, 1973.

Keenleyside, A.: Skeletal evidence of health and disease in pre-contact Alaskan Eskimos and Aleuts. *American Journal of Physical Anthropology 107*(1):51–70, 1998.

Keim, H. A.: Scoliosis. *Clinical Symposia 24,* No. 1, 1972.

___: Low back pain. *Clinical Symposia 25,* No. 3, 1973.

Kelley, M. A.: Parturition and pelvic changes. *American Journal of Physical Anthropology 51*(4):541–546, 1979.

Kelley, M. A., and Eisenberg, L. E.: Blastomycosis and tuberculosis in early American Indians: A biocultural view. *Midcontinental Journal of Archaeology 12*:89–116, 1987.

Kelley, M. A., and El-Najjar, M. V.: Natural variation and differential diagnosis of skeletal changes in tuberculosis. *American Journal of Physical Anthropology 52*:153–167, 1980.

Kelley, M. A., and Micozzi, M. S.: Rib lesions in chronic pulmonary tuberculosis. *American Journal of Physical Anthropology 65*:381–386, 1984.

Kellgren, J. J., and Lawrence, J. S.: Osteoarthritis and disk degeneration in an urban population. *Annals of the Rheumatic Diseases 17*:388–397, 1958.

Kempson, F. C.: Emargination of the patella. *Journal of Anatomy and Physiology 36*:419–420, 1902.

Kennedy, G. E.: The relationship between auditory exostosis and cold water: A latitudinal analysis. *American Journal of Physical Anthropology 71*:401–415, 1986.

Kennedy, K. A. R.: Markers of Occupational Stress: Conspectus and Prognosis of Research. *International Journal of Osteoarchaeology 8*(5):305–310, 1998.

Kessel, L., and Rang, M.: Supracondylar spur of the humerus. *Journal of Bone and Joint Surgery 48-B*(4):765–769, 1966.

Keur, J. J., Campbell, J. P. S., McCarthy, J. F., and Ralph, W. J.: The clinical significance of the elongated styloid process. *Oral Surgery Oral Medicine Oral Pathology 61*:399–404, 1986.

Khazhinskaia, V. A. and Ginzburg, M. A.: X-ray anatomic variants of the rhomboid fossa of the clavicle. *Vestnik Rentgenolog Radiologia 3*:32–37, 1975.

Kim, N, H., and Suk, K. S.: The role of transitional vertebrae in spondylolysis and spondylolytic spondylolisthesis. *Bulletin Hospital Joint Disease 56*:161–166, 1997.

Klepinger, L.L. and Heidingsfelder, J.A.: Probable torticollis revealed in decapitated skull. *Journal of Forensic Sciences 41*(4):693–696, 1996.

Klippel, M., and Feil, A.: Un cas d'absence des vertebres cervicales. *Nouv Iconong Salpetriere 25*:223, 1912.

Klunder, K. B., Rud, B., and Hansen, J.: Osteoarthritis of the hip and knee joint in retired football players. *Acta Orthopaedica Scandinavica 51*:925–927, 1980.

Knight, G. A. M., and Morley, G. H.: Cleft sternum: Case report and brief commentary. *British Journal of Surgery 24*:60–64, 1936–37.

Knowles, A. K.: Acute traumatic lesions. In Hart, G. D. (Ed.): *Disease in ancient man: An international symposium.* Ontario: Irwin, 1983.

Kono, S., Hayashi, N., Kashahara, G.: A study on the etiology of spondylolysis with reference to athletic activities. *Journal of the Japanese Orthopedic Association 49*:125, 1975.

Kortesis, B., Richards, T., David, L., Glazier, S., and Argenta, L.: Surgical management of foramina parietalia permagna. *Journal of Craniofacial Surgery 14*(4):538–544, 2003.

Kostick, E. L.: Facets and imprints on the upper and lower extremities of femora from a Western Nigerian population. *Journal of Anatomy 97*(3):393–402, 1963.

Kramer, S. B., Lee, S. H. S., and Abramson, S. B.: Nonvertebral infections of the musculoskeletal system by Mycobacterium Tuberculosis. In W. N. Rom, S. M. Garay, and B. R. Bloom (editors). *Tuberculosis,* 2nd ed. Philadelphia: Lippincott Williams and Company, 2004.

Krauss, M. J., Morrissey, A. E., Winn, H. N., Amon, E., and Leet, T. L.: Microcephaly: An epidemiologic analysis. *American Journal of Obstetrics and Gynecology 188*(6):1484–1489; discussion 1489–1490, 2003.

Krishnan, J.: Distal radius fractures in adults. *Orthopedics 25*(2):175–179, 2002.

Krogman, W. M.: The pathologies of pre- and protohistoric man. *Ciba Symposia 2*:432–443, 1940.

Krogman, W. M., and Iscan, M. Y.: *The human skeleton in forensic medicine,* 2nd ed. Springfield: Charles C Thomas, 1986.

Kromberg, J. G., and Jenkins, T.: Common birth defects in South African blacks. *South African Medical Journal 62*:599–602, 1982.

Kroon, D. F., Lawson, M. L., Derkay, C. S., Hoffmann, K., and McCook, J.: Surfer's ear: External auditory exostoses are more prevalent in cold water surfers. *Otolaryngology Head Neck Surgery 126*(5):499–504, 2002.

Kulkarni, V. N., and Mehta, J. M.: Tarsal disintegration (TD) in leprosy. *Leprosy in India*

55:338–370, 1983.

Kullmann, L., and Wouters, H. W.: Neurofibromatosis, gigantism and subperiosteal haematoma: Report of two children with extensive subperiosteal bone formation. *Journal of Bone and Joint Surgery 54–A*:130–138, 1972.

Kumai, T., and Benjamin, M.: Heel spur formation and the subcalcaneal enthesis of the plantar fascia. *Journal of Rheumatology 29*(9):1957–1964, 2002.

Kutilek, S., Baxova A., Bayer, M., Leiska, A., Kozlowski, K.: Foramina parietalia permagna: Report of nine cases in one family. *Journal of Paediatric Child Health 33*(2):168–170, 1997.

Ladisch, S., and Jaffe, E. s.: The histocytoses. In Pizzo, P. A. and Poplak, D. G (Eds.): *Principles of pediatric oncology.* Philadelphia: Lippincott, 1989.

Lallo, J. W., Armelagos, G. J., and Mensforth, R. P.: The role of diet, disease and physiology in the origin of porotic hyperostosis. *Human Biology 48*(3):471–483, 1977.

Lambert, P. M.: Rib Lesions in a prehistoric Puebloan sample from southwestern Colorado. *American Journal of Physical Anthropology 117*:281–292, 2002.

Lamy, C., Bazergui, A., Kraus, H., and Farfan, H. F.: The strength of the neural arch and the etiology of spondylolysis. *Orthopedic Clinics of North America 6*:215–231, 1975.

Lane, N. E., Bloch, D. A., Jones, H. H., Simpson, U., and Fries, J. F.: Osteoarthritis in the hand: A comparison of handedness and hand use. *Journal of Rheumatology 16*(5):637–642, 1989.

Lange, R. H., Lange, T. A., and Rao, B. K.: Correlative radiographic, scintigraphic, and histological evaluation of exostoses. *Journal of Bone and Joint Surgery 66–A*(9):1454–1459, 1984.

Langlais, R. P., Miles, D. A., and Van Dis, M. L.: Elongated and mineralized stylohyoid ligament complex: A proposed classification and report of a case of Eagle's syndrome. *Oral Surgery Oral Medicine Oral Pathology 61*:527–532, 1986.

Lanzkowsky, P.: Radiological features of iron deficiency anemia. *American Journal of Diseases of Children 116*:16–29, 1968.

Larsen, W. J.: *Human embryology,* 3rd ed. Philadelphia: Churchill Livingstone, 2001.

Lasater, K., and Groer, M.: Arthritis. In M. Groer (Ed.) *Advanced pathophysiology: Application to clinical practice.* Philadelphia: J.B. Lippincott Company, pp. 245–265, 2000.

Lastres, J. B. and Cabiese, F.: La *Trepanacion del Craneo en el Antiguo Peru.* Lima:Imprenta de la Universidad Nacional Mayor de San Marcos, 1960.

Latham, R. A., and Burston, W. R.: The postnatal pattern of growth at the sutures of the human skull. A histological survey. *The Dental Practitioner and Dental Record 17*(2):61–67, 1966.

Laughlin, W.S. and Jorgensen, J.B.: Isolate variation in Greenlandic Eskimo crania. *Acta Genetica et Statistica Medica 6*:3–12, 1956.

Laurent, L. E., and Einola, S.: Spondylolisthesis in children and adolescents. *Acta Orthopaedica Scandinavica 31*:45–64, 1961.

Lawrence, J. S.: Rheumatism in cotton cooperative. *British Journal Indian Medicine 18*:270–276, 1961.

____: Generalized osteoarthritis in a population sample. *American Journal Epidemiology 90*:381–389, 1969.

____: Relationship of infection to rheumatoid factor in the population. *Annals of Rheumatic Diseases 29*(2):196–197, 1970a.

____: Paget's disease in population samples. *Annals of Rheumatic Diseases 29*(5):562, 1970b.

Lawson, B. J.: Hutchinson's teeth. *Oral Surgery Oral Medicine Oral Pathology 24*(5):635–636, 1967.

Leak, R. S., Rayan, G. M., and Arthur, R. E.: Longitudinal radiographic analysis of rheumatoid arthritis in the hand and wrist. *Journal of Hand Surgery [Am] 28*(3):427–434, 2003.

Leboeuf, C., Kimber, D., and White, K.: Prevalence of spondylolisthesis, transitional anomalies and low intercrestal line in a chiropractic patient population. *Journal of Manipulative Physiology Therapy 12*(3):200–204, 1989.

Lee, T. M., and Mehlman, C. T.: Hyphenated history: Park-Harris growth arrest lines. *American Journal of Orthopedics 32*(8):408–411, 2003.

Lehman, T. J.: Enthesitis, arthritis and heel pain. *Journal of the American Podiatric Medical Association 89*(1):18–19, 1999.

Lepow, G. M., and Cafiero, J. M.: Heterotopic bone formation: A case report. *Journal of Foot Surgery 19*(2):63–65, 1980.

Lester, C. W., and Shapiro, H. L.: Vertebral arch defects in the lumbar vertebrae of pre-historic American Eskimos: A study of skeletons in the American Museum of Natural History, chiefly from Point Hope, Alaska. *American Journal of Physical Anthropology 28*:43–47, 1968.

Lestini, W. F., and Wiesel, S. W.: The pathogenesis of cervical spondylosis. *Clinical Orthopaedics and Related Research 239*:69–93, 1989.

Letts, M., Smallman, T., Afanasiev, R., and Gouw, G.: Fracture of the pars inter-articularis in adolescent athletes: a clinical-biomechanical analysis. *Journal of Pediatric Orthopedics 6*:40–46, 1986.

Leung, J. S. M., Mok, C. K., Leong, J. C. Y., and Chan, W. C.: Syphilitic aortic aneurysm with spinal erosion: Treatment by aneurysm replacement and anterior spinal fusion. *Journal of Bone and Joint Surgery 59-B*(1):89–92, 1977.

Levine, S. M., Lambiase, R. E., and Petchprapa, C. N.: Cortical lesions of the tibia: Characteristic appearances at conventional radiography. *Radiographics 23*(1):157–177, 2003.

Liberson, F.: Os acromiale–A contested anomaly. *Journal of Bone and Joint Surgery 19*:683–689, 1937.

Libson, E., Bloom, R. A., and Dinari, G.: Symptomatic and asymptomatic spondylolysis and spondylolisthesis in young adults. *International Orthopedics 6*:259–261, 1982.

Lichenstein, L.: Histiocytosis X. Integration of eosinophilic granuloma of bone, letterer-Siwe disease, and Schuller-Christian disease as related manifestations of a single nosologic entity. *Archives of Pathology 56*:84, 1953.

____: *Diseases of bone and joints.* Saint Louis: Mosby, 1970.

Lidov, M., and Som P. M.: Inflammatory disease involving a concha bullosa (enlarged pneumatized middle nasal turbinate): MR and CT appearance. *American Journal Neuroradiology 11*(5):999–1001, 1990.

Lima, Mauricio D. L. P.: *Contribucicao ao Estudo do Os Trigonum Tarsi.* San Paulo, These Inaugural, 1928.

Limson, M.: Metopism as found in Filipino skulls. *American Journal of Physical Anthropology 7*(3):317–324, 1924.

Lindahl, S., Nyman, R. S., Brismar, J., Hugosson, C., and Lundstedt, C.: Imaging of tuberculosis. IV. Spinal manifestation in 63 patients. *Acta Radiologica 37*:506–511, 1996.

Lindau, T. R., Aspenberg, P., Arner, M., Redlundh-Johnell, I., and Hagberg, L.: Fractures of the distal forearm in young adults. An epidemiologic description of 341 patients. *Acta Orthopaedica Scandinavica 70*(2):124–128, 1999.

Lindblom, K.: Bachache and its relation to ruptures of the intravertebral disks. *Radiology 57*:710, 1951.

Lisowski, F. P.: Prehistoric and early historic trepanation. In Brothwell, D. R. and Sandison, A. T. (Eds.): *Diseases in antiquity: A survey of diseases injuries and surgery of early populations.* Springfield: Thomas, 1967.

Livingstone, F.: On the origin of syphilis: An alternative hypothesis. *Current Anthropology 32*:587–590, 1991.

Lodge, T.: Thinning of the parietal bones in early Egyptian populations and its aetiology in the light of modern observations. In Brothwell, D. R. and Sandison, A. T. (Eds.): *Diseases in antiquity: A survey of diseases injuries and surgery of early populations.* Springfield: Thomas, 1967.

Lombardi, C. M., Silhanek, A. D., and Connolly, F. G.: Modified arthroscopic excision of the symptomatic os trigonum and release of the flexor hallucis longus tendon: Operative technique and case study. *Journal of Foot and Ankle Surgery, 5,* Vol. 38: 347–351, September/October, 1999.

Long, B. W. and Rafert, J. A.: *Orthopaedic radiography.* Philadelphia: W. B. Saunders Company, 1995.

Longia, G. S., Agarwal, A. K., Thomas, R. J., Jain, P. N., and Saxena, S, K.: Metrical study of rhomboid fossa of clavicle. *Anthropologischer Anzeiger 40*(2):111–115, 1982.

Longridge, N. S.: Exostosis of the external auditory canal: A technical note. *Otology Neurology 23*(3):260–261, 2002.

Lowe, J., Libson, J., Ziv, I., Nyska, M., Floman, Y., Bloom, and Robin, G. C.: Spondylolysis in the upper lumbar spine. A study of 32 patients. *Journal of Bone and Joint Surgery 39-B*(69):582–586, 1987.

Lu, H., Gu, G., and Zhu, S.: Heel pain and calcaneal spurs (article in Chinese, English abstract). *Zhonghua Wai Ke Za Zhi 34*(5):294–296, 1996.

Lucas, M. F.: (1) Two cases of cervical ribs. (2) An anomalous arrangement of the vagi. *Journal of Anatomy and Physiology 69*:336–342, 1915.

Lundy, J. K.: A report on the use of Fully's anatomical method to estimate stature in military skeletal remains. *Journal of Forensic Sciences 33*(2):534, 1988.

Luong, A. A., and Salonen, D. C.: Imaging of the seronegative spondyloarthropathies. *Current Rheumatology Preparation 2*(4):288–296, 2000.

Ly, J. Q., Sanders, T. G., Mulloy, J. P., Soares, G. M., Beall, D. P., Parsons, T. W., and Slabaugh, M. A.: Osseous change adjacent to soft-tissue hemangiomas of the extremities: Correlation with lesion size and proximity to bone. *American Journal Roentgenology 180*(6):1695–1700, 2003.

Maas, M., Slim, E. J., Heoksma, A. F., van der Kleij, A. J., Akkerman, E. M., den Heeten G. J., and Faber, W. R.: MR imaging of neuropathic feet in leprosy patients with suspected osteomyelitis. *International Journal Leprosy Other Mycobacterial Disease 70*(2):97–103, 2002.

MacAusland, W. R., and Mayo, R. A.: *Orthopedics: A concise guide to clinical practices.* Boston: Little, Brown, 1965.

MacInnis, E. L., Hardie, J., Baig, M., and al-Sanea, R. A.: Gigantiform Torus palatinus: Review of the literature and report of a case. *International Dental Journal 48*(1):40–43, 1998.

Madewell, J. E., Ragsdale, B. D., and Sweet, D. E.: Radiologic and pathologic analysis of solitary bone lesions. Part I: internal margins. *Radiologic Clinics of North America 19*(4):715–748, 1981a.

____: Radiologic and pathologic analysis of solitary bone lesions. Part II: periosteal reactions. *Radiologic Clinics of North America 19*(4):749–783, 1981b.

Magee, D. J.: *Orthopedic physical assessment,* 4th ed. Philadelphia: Saunders, 2002.

Manchester, K.: *The archaeology of disease.* Bradford, University of Bradford, 1983.

____: Bone changes in leprosy: Pathogenesis and paleopathology, Paper presented at the Annual Meeting of the Association of Physical Anthropologists, San Diego, 1989.

Mankin, H. J.: The structure, chemistry and metabolism of articular cartilage. *Bulletin of Rheumatic Diseases 17*(7):447–452, 1967.

____: The effect of aging on articular cartilage. *Bulletin of the New York Academy of Medicine 44*(5):545–552, 1968.

____: The reaction of articular cartilage to injury and osteoarthritis (second of two parts). *New England Journal of Medicine 291*(25, Dec. 19):1335–40, 1974.

____: Rickets, osteomalacia, and renal osteodystrophy, part I. *Journal of Bone and Joint Surgery 56-A*:101, 1974.

____: Rickets, osteomalacia, and renal osteodystrophy, part II. *Journal of Bone and Joint Surgery 56-B*:352–386, 1974.

____: The response of articular cartilage to mechanical injury. *Journal of Bone and Joint Surgery 64*(3):460–466, 1982.

____: Rickets, osteomalacia, and renal osteodystrophy. An update. *Orthopedic Clinics of North America 21*(1):81–96, 1990.

Mann, R. W.: *Stafne's defect of the human mandible.* Doctoral Dissertation, University of Hawaii, Manoa, 2001.

____: Calcaneus secundarius: Description of a common accessory ossicle. *Journal of American Podiatric Medical Association 79*(8):363–366, 1989.

____: A method for siding and sequencing human ribs. *Journal of Forensic Sciences 38*(1):151–155,

1993.

___: The Calcaneus Secundarius: Frequency and description of an accessory ossicle in six samples. *American Journal of Physical Anthropology 81:*17–25, 1990.

___: Enlarged parietal foramina and craniosynostosis in an American Indian child. *American Journal of Roentgenology 154:*658, 1990.

Mann, R. W., and Murphy, S. P.: Skeletal indicators of physical stress in soldiers from the Battle of Fort Erie (War of 1812), Canada. Paper presented at the Annual Meeting of the Northeastern Anthropological Association, Montreal, Canada, 1989.

Mann, R. W., and Murphy, S. P.: Skeletal indicators of stress in 28 American soldiers who died in the Battle of Fort Erie, War of 1812. *Northeastern Anthropology Association 26,* 1989.

Mann, R. W., and Owsley, D. W.: Anatomy of uncorrected talipes equinovarus in a fifteenth-century Ameican Indian. *Journal of the American Podiatric Medical Association 79*(9):436–440, 1989.

Mann, R. W., and Owsley, D. W.: Os Trigonum: Variation of a common accessory ossicle of the talus. *Journal of the American Podiatric Medical Association 80*(11):623–625, 1990.

Mann, R. W., and Owsley, D. W.: Human osteology: Key to the sequence of events in a post-mortem shooting. *Journal of Forensic Sciences 37*(5):1386–1392, 1992.

Mann, R. W., and Verano, J. W.: Congenital spinal anomalies in a prehistoric adult female from Peru. *Paleopathology Newsletter 72,* 1990.

Mann, R. W., Meadows, L., Bass, W. M., and Watters, D. R.: Description of skeletal remains from a black slave cemetery from Montserrat, West Indies. *Annals of Carnegie Museum 56:*319–336, 1987.

Mann, R. W., Owsley, D. W., and Sledzik, P. S.: Seronegative spondyloarthropathy of the foot. *Journal of the American Podiatric Medical Association 80*(7):360–363, 1990.

Mann, R. W., Sledzik, P. S., Owsley, D. W., and Droulette, M. R.: Radiographic examination of Chinese foot binding. *Journal of the American Podiatric Medical Association 80*(8):405–409, 1990.

Manzanares, M. C., Goret-Nicaise, M., and Dhem, A.: Metopic sutrual closure in the human skull. *Journal of Anatomy 161:*203–215, 1988.

Margetts, E. L.: Trepanation of the skull by the Medicine-men of primative cultures, with particular reference to present day native east African practice. In Brothwell, D. R. and Sandison, A. T. (Eds.): *Diseases in antiquity: A survey of the diseases injuries and surgery of early populations.* Springfield: Thomas, pp. 673–701, 1967.

Markowitz, H. A., and Gerry, R. G.: Temporomandibular joint disease. *Oral Surgery, Oral Medicine, Oral Pathology 3:*75, 1950.

Marotta, J. J., and Micheli, L, J.: Os trigonum impingement in dancers. *American Journal of Sports Medicine 20*(5):533–536, 1992.

Marquis, J. W., Bruwer, A. J., and Kieth, H. M.: Supracondyloid process of the humerus. *Mayo Clinic Proceedings 32*(24, Nov. 27):691–697, 1957.

Martin, D. S., and Smith D.T.: Blastomycosis. *American Revue of Tuberculosis 39:*275–304, 1939.

Masciocchi, C., Catalucci, A. and Barile, A.: Ankle impingement syndromes. *European Journal of Radiology 27* Suppl 1:S70–73, 1998.

Masi, A. T., and Medsger, T. A.: Epidemiology of the rheumatic diseases. In McCarty, D. J. (Ed.): *Arthritis and allied conditions: A textbook of rheumatology,* 11th ed. Philadelphia: Lea and Febiger, 1989.

Massada, J. L.: Ankle overuse in soccer players: Morphological adaptation of the talus in the anterior impingement. *Journal of Sports Medicine and Physical Fitness 31*(3):447–451, 1991.

Matsumura, G., Uchiumi, T., Kida, K., Ichikawa, R., and Kodama, G.: Developmental studies on the interparietal part of the human occipital squama. *Journal Anatomy 182*(Pt 2):197–204, 1993.

Mays, S.: *The archaeology of human bones.* London: Routledge Heritage, 1998.

Mays, S., Crane-Kramer, G., and Bayliss, A.: Two Probable cases of Treponemal Disease of Medieval Date From England. *American Journal of Physical Anthropology 120:*133–143, 2003.

Mays, S., Strouhal, E., Vyhnanek, L., and Nemeckova, A.: A case of metastatic carcinoma of medieval date from Wharram Percy, England. *Journal of Paleopathology 8*:33–42, 1996.

Mays, S., Taylor, G. M., Legge, A. J., Young, D. B., and Turner-Walker, G.: Paleopathological and biomolecular study of tuberculosis in a medieval skeletal collection from England. *American Journal of Physical Anthropology 114*:298–311, 2001.

McCarty, D. J. (Ed.): *Arthritis and allied conditions: A textbook of rheumatology,* 11th ed. Philadelphia: Lea and Febiger, 1989.

McCormick, W.: Sternal foramina in man. *American Journal of Forensic Medicine and Pathology 2*:249–252, 1981.

McCort, J. J. and Mindelzun, R. E.: *Trauma radiology.* New York: Churchill Livingstone, 1990.

McDougall, A.: The os trigonum. *Journal of Bone and Joint Surgery 37-B*:257–265, 1955.

McKee, B. W., Alexander, W. J., and Dinbar, J. S.: Spondylolysis and spondylolistesis in children. *Journal of the Canadian Association of Radiology 22*:100, 1971.

McKern, T. W., and Stewart, T. D.: *Skeletal age changes in young American males: Analyzed from the standpoint of age identification.* Technical Report EP-45, Quartermaster Research and Development Center, Massachusetts, 1957.

McKusick, V. A.: *Human genetics,* 2nd ed. Englewood Cliffs: Prentice Hall, 1969.

Meisel, A. D., and Bullough, P. G.: *Atlas of osteoarthritis.* New York: Lea and Febiger, 1984.

Mensforth, R. P., Lovejoy, C. O., Lallo, J. W., and Armelagos, G. J.: The role of constitutional factors, diet, and infectious disease in the etiology of porotic hyperostosis and periosteal reactions in prehistoric infants and children. *Medical Anthropology 2* (Winter), part 2, pp. 1–59, 1978.

Merbs, C. F.: *Patterns of activity-induced pathology in a Canadian Inuit population.* Archaeological Survey of Canada Paper, Mercury Series 119. Ottawa: National Museum of Man, 1983.

Merbs, C. F.: Incomplete spondylolysis and healing: A study of ancient Canadian Eskimo skeletons. *Spine 20*:2328–2334, 1995.

Merbs, C. F.: Spondylolysis of the sacrum in Alaskan and Canadian Inuit skeletons. *American Journal of Physical Anthropology 101*:357–367, 1996.

Merbs, C. F.: Degenerative spondylolisthesis in ancient and historic skeletons from New Mexico Pueblo sites. *American Journal of Physical Anthropology 116*:285–295.

Merbs, C. F.: Asymmetrical Spondylolysis. *American Journal of Physical Anthropology 119*:156–174, 2002.

Merbs, C. F., and Euler, R. C.: Atlanto-occipital fusion and spondylolisthesis in an Anasazi skeleton from Bright Angle Ruin, Grand Canyon National Park, Arizona. *American Journal of Physical Anthropology 67*:381–391, 1985.

Meschan, I.: *Roentgen signs in diagnostic imaging* (2nd edition). Philadelphia: W. B. Saunders, 1984.

Meyer, A. W.: The "Cervical Fossa" or Allen. *American Journal of Physical Anthropology 7*(2):257–269, 1924.

Micheli, L. J., Slater, J. A., Woods, E., and Gerbino, P. G. (1986): Patella alta and the adolescent growth spurt. *Clinical Orthopedics 213*:159–162.

Micozzi, M. S., and Kelley, M. A.: Evidence for precolumbian tuberculosis at the Point of Pines Site, Arizona: Skeletal pathology in the sacro-iliac region. In Merbs, C. F. and Miller, R. J. (Eds.): *Health and Disease in the Prehistoric Southwest.* Anthropological Research Papers No. 34. Tempe, Arizona State University Press, 1985.

Mihran, O., and Tachdjian, M. D.: *Atlas of Pediatric Orthopaedic Surgery.* Philadelphia: W. B. Saunders, 1994.

Mii, S., Mori, A., Yamaoka, T., and Sakata, H.: Penetration by a huge abdominal aortic aneurysm into the lumbar vertebrae: Report of a case. *Surgery Today 29*(12):1299–1300, 1999.

Miki, T., Tamura, T., Senzoku, F., Kotani, H., Hara, T., and Masuda, T.: Congenital laminar defect of the upper lumbar spine associated with pars defect. A report of eleven cases. *Spine 16*:353–355, 1991.

Miles, J. S.: *Orthopedic Problems of the Wetherill Mesa Populations.* Washington, D.C.: Publications in Archaeology 7G. National Park Service, 1975.

Mintz, G., and Fraga, A.: Severe osteoarthritis of the elbow in foundry workers. *Archives Environment Health 27*:78–80, 1973.

Mirra, J. M.: Pathogenesis of Paget's disease based on viral etiology. *Clinical Orthopaedics and Related Research 217*:162–170, 1987.

Mirra, J. M., Brien, E. W., and Tehranzadeh, J.: Paget's disease of bone: review with emphasis on radiologic features, part I. *Skeletal Radiology 24*:163–171, 1995 (a).

___: Paget's disease of bone: Review with emphasis on radiologic features, part II. *Skeletal Radiology 24*:173–184, 1995 (b).

Mittler, D. M., and Van Gerven D. P.: Developmental, diachronic, and demographic analysis of cribra orbitalia in the medieval Christian populations of Kulubnarti. *American Journal of Physical Anthropology 93*(3): 287–297, 1994.

Moe, J. H., Winter, R. B., Bradford, D. S., and Lonstein, J. E.: *Scoliosis and other spinal deformities.* Philadelphia: Saunders, 1978.

Møller-Christensen, V.: *Ten lepers from Naestved in Denmark: A study of skeletons from a medieval Danish leper hospital.* Copenhagen: Danish Science Press, 1953.

___: *Bone changes in leprosy.* Copenhagen: Munksgaard, 1961.

___: Evidence of leprosy in earlier peoples. In Brothwell, D. R. and Sandison, A. T. (Eds.): *Diseases in antiquity: A survey of the diseases injuries and surgery of early populations.* Springfield: Thomas, pp. 295–306, 1967.

___: Changes in the anterior nasal spine and alveolar process of maxilla in leprosy: A clinical examination. *International Journal of Leprosy 42*:431–435, 1974.

___: *Leprosy changes in the skull.* Odense: Odense University Press, 1978.

___: Leprosy and tuberculosis. In Hart, G. D. (Ed.): *Disease in ancient man: An international symposium.* Ontario: Irwin, pp. 129–138, 1983.

Møller-Christensen, V., and Sandison, A. T.: Ursa orbitae (cribra orbitalia) in the collection of crania in the anatomy department of the University of Glascow. *Pathologia et Microbiologia 26*:175–183, 1963.

Molleson T., P. Andrews, B. Boz, J. Sofaer Derevenski and J. Pearson. *Human remains up to 1998. Catalhoyuk 1998 Archive Report* (catal.arch.cam.ac.uk/index.html).

Molto, J.E.: *Biological relationships of Southern Ontario Woodland peoples: the evidence of discontinuous cranial morphology.* National Museum of Man (Mercury series), Archeological Survey of Canada, Paper 117, 1983.

Monsour, P. A., and Young, W. G.: Variability of the styloid process and stylohyoid ligament in panoramic radiographs. *Oral Surgery Oral Medicine Oral Pathology 61*:522–526, 1986.

Moore, K. L.: *The developing human: Clinically oriented embryology.* London: Saunders, 1982.

Moore, K. L., and Persaud, T. V. N.: *Before we are born: Esentials of embryology and birth eefects,* 6th edition. Philadelphia: Saunders, 2003.

Moore, R. M., and Green, N. E.: Blastomycosis of bone: a report of six cases. *Journal of Bone and Joint Surgery 64-A*(7):1097–1101, 1982.

Moore, S.: *Hyperostosis cranii.* Springfield: Thomas, 1955.

Moore, S. L., and Rafii, M.: Imaging of musculoskeletal and spinal tuberculosis. *Radiologic Clinics of North American 39*:329–342, 2001.

Moore, T. E., Kathol, M. H., El-Khoury, G. Y., Walker, C. W., Gendall, P. W., and Whitten, C. G.: Unusual radiological features in Paget's disease of bone. *Skeletal Radiology 23*:257–260, 1994.

Moore, T. E., King, A. R., Kathol, M. H., El-Khoury, G. Y., Palmer, R., and Downey, P. R.: Sarcoma in Paget disease of bone: Clinical, radiologic and pathologic features in 22 cases. *American Journal Radiology 156*:1199–1203, 1990.

Morrissy, R. T., Weinstein, S. L., and Kida, B.: *Atlas of Pediatric Orthopaedic Surgery.* Philadelphia: J. B. Lippincott, 2000.

Morse, D.: Prehistoric tuberculosis in America. *American Review of Respiratory Disease 83*:489–504, 1961.

___: *Ancient disease in the midwest.* Illinois State Museum Reports of Investigations, No. 15, 1969.

Morton, N. E.: Birth defects in racial crosses. In *Congenital Malformation: Proceedings of the Third International Conference.* New York: Excerpta Medica, 1970.

Moseley, J. E.: *Bone changes in hematologic disorders.* New York: Grune and Stratton, 1963.

Moskowitz, R. W., Howell, D. S., Goldberg, V. M., and Mankin, H. J.: *Osteoarthritis: Diagnosis and management.* Philadelphia: Saunders, 1984.

Moss, M.L.: The pathogenesis of artificial cranial deformation. *American Journal of Physical Anthropology 16:*269–86, 1958.

Muckley, T., Schutz, T. Kirschner, M., Potulski, M., Hofmann, G., and Buhren, V.: Psoas abscess: The spine as a primary source of infection. *Spine 28*(6):E106–113, 2003.

Mudge, M. K., Wood, V. E., and Frykman, G. K.: Rotator cuff tears associated with os acromiale. *Journal of Bone and Joint Surgery 66-A*(3):427–429, 1984.

Muller, F., O'Rahilly, R., and Benson, D. R.: The early origin of vertebral anomalies, as illustrated by a "butterfly vertebra." *Journal of Anatomy 149:*157–169, 1986.

Murphy, J. and Gooding, C.A.: Evolution of persistently enlarged parietal foramina. *Radiology 97:*391–392, 1970.

Murphy, S. P., and Mann, R. W.: Cortical defects of the proximal humerus: An indicator of physical stress. *American Journal of Physical Anthropology 81*(2):273, 1990.

Murray, J. F., Merriweather, A. M., and Freedman, M. L.: Endemic syphilis in the Bakwena Reserve of the Bechuanaland protectorate. *Bulletin of World Health Organization 15:*975–1039, 1956.

Murray, K.: Pers. Comm., Smithsonian Institution, Washington, DC, 1989.

Murray-Leslie, D. F., Lintott, D. J., and Wright, V.: The knees and ankles in sport and veteran military parachutists. *Annals Rheumatic Disease 36:*327–337, 1977.

Myers, W. J.: Anterior ankle impingement exostoses. *Journal of American Podiatric Medical Association 77*(7):347–350, 1987.

Naffsiger, H. C., Inman, V., and Saunders, J.: Lesions of intervertebral discand ligamenta flava: clinical and anatomical studies. *Surgery, Gynecology and Obstetrics 66:*288, 1938.

Nathan, H.: Spondylolysis: Its anatomy and mechanism of development. *Journal of Bone and Joint Surgery 41-A:*303–320, 1959.

Nelsen, K., Tayles, N., and Domett, K.: Missing lateral incisors in Iron Age South-East Asians as possible indicators of dental agenesis. *Archives Oral Biology 46*(10):963–971, 2001.

Neugebauer, F. L.: The classic: A new contribution to the history and etiology of spondylolisthesis. In Urist, M. R. (editor-in-Chief): *Clinical Orthopaedics and Related Research* (No. 107). Philadephia: J. B. Lippincott Company, pp. 4–22, 1976.

Neumann, G. K.: Evidence for the antiquity of scalping from Central Illinois. *American Antiquity 5:*287–289, 1940.

Newman, P. H. and Stone K.H.: The etiology of spondylolysis. *Journal of Bone and Joint Surgery 45-B*(1):39–59, 1963.

Nielsen, S. P., Xie, X., and Barenholdt, O.: Geometric properties of distal radius and pathogenesis of Colles fracture: a peripheral quantitative computed tomography study. *Journal of Clinical Densitometry 4*(3):209–219, 2001.

Niswander, J. D., Barrow, M. V., and Bingle, G. J.: Congenital malformations in the American Indian. *Social Biology 22*(3):203–215, 1975.

Norman, A.: Roentgenologic diagnosis. In Moskowitz, R. W., Howell, D. S., Goldberg, V. M., and Mankin, H. J. (Eds.): *Osteoarthritis: Diagnosis and management.* Philadelphia: Saunders, 1984.

Norman, A. P.: *Congenital abnormalities in infancy.* Philadelphia: F.A. Davis, 1963.

Nowak, O., and Piontek, J.: The frequency of appearance of transverse (Harris) lines in The tibia in relationship to age at death. *Annals of Human Biology 29*(3):314–325, 2002 (a).

___: Does the occurrence of Harris lines affect the morphology of human long bones? *Homo 52*(3):254–276, 2002 (b).

Nugent, C. A., Gall, E. P., and Pitt, M. J.: Osteoporosis, osteomalacia, rickets and Paget's disease. *Primary Care 11*(2):353–368, 1984.

Nyland, H., and Krogness, K. G.: The normal calvaria indices of the human skull. 1. Absolute measurements. 2. Proportional measurements. *Pediatric Radiology* 7(1):1–3, 1978.

O'Beirne, J. G., and Horgan, J. G.: Stress fracture of the lamina associated with unilateral spondylosysis. *Spine 13:*220–222, 1988.

Oettenking, B.: Anomalous patellae. *Anatomical Record 23:*269–279, 1922.

Ogden, J. A., McCarthy, S. M., and Jokl, P.: The painful bipartite patella. *Journal of Pediatric Orthopedics* 2(3):263–269, 1982.

Ohmori, K. Yoshihiro Ishida, Takatsu, T., Inque, H., Suzuki, K.: Vertebral slip in lumbar spondylolysis and spondylolisthesis: Long-term follow-up of 22 adult patients. *Journal of Bone and Joint Surgery 77–B*(5), 771–773, 1995.

Oloff, L. M., Schulhofer, S. D., and Bocko, A. P.: Subtalar joint arthroscopy for sinus tarsi syndrome: A review of 29 cases. *Journal of Foot and Ankle Surgery 40*(3):152–157, 2001.

Olmsted, W. W.: Some skeletogenic lesions with common calvarial manifestations. *Radiologic Clinics of North America 19:*703–713, 1981.

Omey, M. L., Micheli, L. J., and Gerbino, P. G.: Idiopathic scoliosis and spondylolysis in the female athlete. Tips for treatment. *Clinical Orthopedics 372:*74–84, 2000.

Omnell, K. A., Gandhi, C., and Omnell, M. L.: Ossification of the human stylohyoid ligament: A longitudinal study. *Oral Surgery Oral Medicine Oral Pathology Oral Radiology Endodontics 85*(2):226–232, 1998.

O'Neal, M. L., Dwornik, J. J., Ganey, T. M, and Ogden J. A.: Postnatal development of the human sternum. *Journal of Pediatric Orthopedics, 18* (3):398–405, 1998.

O'Rahilly, R. M., and Muller, F.: *Human Embryology and Teratology,* 2nd ed. New York: Wiley-Liss, 1996.

O'Rahilly, R. M., and Twohig, M.: Foramina parietalia permagna. *American Journal of Roentgenology 67:*551–561, 1952.

Ortner, D. J.: Descriptions and classifications of degenerative bone changes in the distal joint surfaces of the humerus. *American Journal of Physical Anthropology 28:*139–155, 1968.

Ortner, D. J.: *Identification of Pathological Conditions in Human Skeletal Remains,* 2nd edition. Amsterdam: Academic Press, 2003.

Ortner, D. J., and Aufderheide, A. C (Eds).: *Human paleopathology: Current syntheses and future options.* Washington, D. C.: Smithsonian Institution Press, 1991.

Ortner, D. J., Butler, W., Cafarella, J, and Milligan, L.: Evidence of probable scurvy in subadults from archeological sites in North America. *American Journal of Physical Anthropology 114:*343–351, 2001.

Ortner, D. J., and Putschar, W. G. J.: *Identification of pathological conditions in human skeletal remains.* Washington: Smithsonian Institution Press, 1985.

Ortner, D.J., and Turner-Walker, G.: The biology of skeletal tissues. In Ortner, D. J.: *Identification of pathological conditions in human skeletal remains,* 2nd edition. Amsterdam: Academic Press, 2003:11–35.

Outerbridge, R. E.: Further studies on the etiology of chondromalacia patellae. *The Journal of Bone and Joint Surgery 46–B,* No. 2: 179–190, 1964.

Owsley, D. W., and Mann, R. W.: An Amercian Indian Skeleton with clubfoot from the Cabin Burial Site (A1184), Hemphill County, Texas. *Plains Anthropologist 35:*93–101, 1990.

Owsley, D. W., Orser, C. E., Mann, R. W., Moore-Jansen, P. H., and Montgomery R. L.: Demography and pathology of an urban slave population from New Orleans. *American Journal of Physical Anthropology 74:*185–197, 1987.

Oygucu, I. H., Kurt, M. A., Ikiz, I., Erem, T., and Davies, D. C.: Squatting facets of the Talus and extensions of the trochlear surface of the talus in late Byzantine males. *Journal of Anatomy 192*(Pt 2):287–291, 1998.

Ozbek, M.: Cranial deformation in a subadult sample from degirmentepe (Chalcolithic, Turkey). *American Journal of Physical Anthropology 115:*238–244, 2001.

Paget, J.: On the production of some of the loose bodies in the joints. *St. Bartholomew's Hospital Report 6:*1, 1870.

____: On a form of chronic inflammation of bones (osteitis deformans). *Transactions of the Medical Chirurgical Society 60:*37–64, 1877.

Pálfi, G., Dutour, O., Deák, J., Hutás, I. (Eds.): *Tuberculosis: Past and present.* Tuberculosis Foundation, Budapest, Szeged and Golden Book, 1999.

Palkovich, A. M.: Endemic disease patterns in paleopathology: Porotic hyperostosis. *American Journal of Physical Anthropology 74:*526–537, 1987.

Pandey, S. K., and Singh, S.: Study of squatting facet/extension of the talus in both sexes. *Medical Science and Law 30*(2):159–164, 1990.

Pang, D., and Lin A.: Symptomatic large parietal foramina. *Neurosurgery 11*(1 Pt 1): 33–37, 1982.

Panush, R. S., Schmidt, C., Caldwell, J. R., Edwards, N. L. Longley, S., Yonker, R., Webster, E., Nauman, J., and Pettersson., H.: Is running associated with degenerative joint disease? *Journal of American Medical Association 255:*1152–1154, 1986.

Papadopoulos, C. C.: Temporal variatin and sex differences in the incidence of cranial porotic hyperostosis in Peru. *Paleopathology Newsletter 19:*11–14, 1977.

Paparella, M. M., Goycoolea, M. V., and Meyerhoff, W. L.: Inner ear pathology and otitis media: A review. *Annals of Otology Rhinology and Laryngology 89*(68):249–253, 1980.

Park, E. A.: The imprinting of nutritional disturbances on the growing bone. *Pediatrics 33:*815–862, 1964.

Park, J.G., Lee, J.K. and Phelps, C.T.: Os acromiale associated with rotator cuff impingement: MR Imaging of the shoulder. *Musculoskeletal Radiology 193:*255–257, 1994.

Parkes, J. C. II, Hamilton, W. G. Patterson, A. J., Rawles, J. G. Jr.: The anterior impingement syndrome of the ankle. *Journal of Trauma 20*(10):895–898, 1980.

Parkinson, C. E.: The supracondyloid process. *Radiology 62:*556–558, 1954.

Parsons, F. G.: On the proportions and characteristics of the modern English clavicle. *Journal of Anatomy 51:*71, 1916.

Patankar, T., Krishnan, A., Patkar, D., Kale, H., Prasad, S., Shah, J., and Castillo, M.: Imaging in isolated sacral tuberculosis: A review of 15 cases. *Skeletal Radiology 29:*392–396, 2000.

Paterson, D. E.: Bones changes in Leprosy. *Leprosy in India 28:*128, 1965.

Paterson, D. E., and Job, C. K.: Bone changes and absorption in leprosy. In Cochrane, R. G. and T. F. Davey (Eds.): *Leprosy in theory and practice.* Bristol, John Wright and Sons, 1964.

Patil, K. M., and Jacob, S.: Mechanics of tarsal disintegration and plantar ulcers in leprosy by stress analysis in three dimensional foot models. *Indian Journal of Leprosy 72*(1):69–86, 2000.

Patni, V. M., Gadewar, D. R., and Pillai, K. G.: Ossification of stylohyoid ligament with pseudojoint formation–A case report. *Journal of Indian Dental Association 58*(6):227–231, 1986.

Pavithran, K.: Saber tibiae in lepromatous leprosy. *International Journal of Leprosy and Other Mycobaterial Disorders 58*(2):385–387, 1990.

Pecina, M. M., and Bojanic, I.: *Overuse Injuries of the Musculoskeletal System,* 2nd ed. Boca Raton, CRC Press, 2004.

Pedersen, A.K., and Hagen, R.: Spondylolysis and Spondylolisthesis. *Journal of Bone and Joint Surgery 70–A*(1):15–24, 1988.

Pepper, O.H.P. and Pendergrass, E.P.: Heredity occurance of enlarged parietal foramina. *American Journal of Roentgenology 35*(1):1–8, 1936.

Percy, E. C., and Mann, D. L.: Tarsal coalition: a review of the literature and presentation of 13 cases. *Foot and Ankle 9*(1):40–44, 1988.

Peterson, J.: The Natufian Hunting conundrum: Spears, atlatls, or bows? Musculoskeletal and armature evidence. *International Journal of Osteoarchaeology 8:*378–389, 1998.

Perou, M.: *Cranial hyperostosis.* Springfield: C. C Thomas, 1964.

Pfeiffer, S.: Morbidity and Mortality in the Uxbridge Ossuary. *Canadian Review of Physical Anthropology 4*(2):23–31, 1985.

Pfeiffer, S.: Rib lesions and New World tuberculosis. *International Journal of Osteoarchaeology 1:*191–198. 1991.

Pindborg, J. J.: *Pathology of the dental hard tissue.* Philadelphia: Saunders, 1970.

Pitt, M. J.: Rachitic and osteomalacic syndromes. *Radiologic Clinics of North America 19*(4):581–599, 1981.

Pitt, M. J.: Rickets and osteomalacia are still around. *Radiological Clinics of North America* 29(1):97–118, 1991.

Plotz G. M., Rrymka, M., Knoch, M. V., Markova, B., and Ulrich, H. W.: Spontaneous fusion of "os acetabuli" after triple pelvic osteotomy. *Archives Orthopaedics Trauma Surgery,* 122(9–10):526–529, 2002.

Poirier, P.: *Traite' d'anatomie humaine I:*515. Poireir & Charpy, Paris, 1911.

Ponec, D. J., and Resnick, D.: On the etiology and pathogenesis of porotic hyperostosis of the skull. *Investigative Radiology* 19(4):313–317, 1984.

Ponseti, I. V.: Growth and development of the acetabulum in the normal child: Anatomical, histological, and roentgenographic studies. *Journal of Bone and Joint Surg. A,*60(5): 575–585, 1978.

___: Treatment of congenital clubfoot. *Journal of Orthopaedics Sports Physical Therapy* 20(1):1, 1994.

Ponseti, I. V., El-Khoury, G. Y., Ippolito, E., and Weinstein, S. L.: A radiographic study of skeletal deformities in treated clubfeet. *Clinical Orthopaedics and Related Research 160:*30–42, 1981.

Popowsky, J.: Bifurcated extremities of the ribs. *Anatomischen Anzeiger 15.*284, 1918.

Porter, R. W. and Park, W.: Unilateral spondylolysis. *Journal of Bone and Joint Surgery* 64-B(3):344–348, 1982.

Posch, T. J., and Puckett, M. L.: Marrow MR signal abnormality associated with bilateral avulsive cortical irregularities in a gymnast. *Skeletal Radiology 27:*511–514, 1998.

Poswall, B. D.: Coccidioidomycosis and North American blastomycosis: Differential diagnosis of bone lesions in pre-Columbian American Indians. *American Journal of Physical Anthropology 44:*199–200, 1976.

Poznanski, A. K.: *The hand and radiologic diagnosis.* Philadelphia: Saunders, 1974.

Prasad, K. C., Kamath, M. P., Reddy, K. J., Raju, K., and Agarwal, S.: Elongated styloid process (Eagle's syndrome): A clinical study. *Journal of Oral Maxillofacial Surgery* 60(2):171–175, 2002.

Procknow, J. J., and Loosli, C. G.: Treatment of the deep mycosis. *American Medical Association Archives of Internal Medicine 101:*765–802, 1958.

Prokopec, M., Simpson, D., Morris, L., and Pretty, G.: Craniosynostosis in a prehistoric aboriginal skull: Case report. *OSSA 9/11:*111–118, 1982–1984.

Prusick, V. R., Samberg, L. C., and Wesolowski, D. P.: Klippel-Feil syndrome associated with spinal stenosis: A case report. *Journal of Bone and Joint Surgery 67-A:*161–164, 1985.

Pulec, J. L., and Deguine, C.: Osteoma of the external auditory canal. *Ear Nose and Throat Journal 79*(12):908, 2000.

___: Nonobstructing exostoses of the auditory canal. *Ear Nose and Throat Journal 80*(2):66, 2001.

___: Exostoses of the external auditory canal. *Ear Nose and Throat Journal 80*(4):190, 2001.

Puranen, J., Ala-Ketola, L., Peltokallio, P., and Saarela, J.: Running and primary osteoarthritis of the hip. *British Medical Journal 2:*424–425, 1975.

Purves, R. K., and Wedin, P. H.: Familial incidence of cervical ribs. *Journal of Thoracic Surgery 19.*952–956, 1950.

Pusey, W. A.: the beginning of syphilis. *Journal of American Medical Association 44:*1961–1964, 1915.

Pynn, B. R., Kurys-Kos, N. S., Walker, D. A., and Mayhall, J. T.: Tori mandibularis: a case report and review of the literature. *Journal Canadian Dental Association 61*(12):1957–1058, 1063–1066, 1995.

Quatrehomme, G., and Iscan, Y. M.: Characteristics of Gunshot Wounds in the Skull. *Journal of Forensic Sciences 44*(3):568–576, 1999.

Queneau, P., Gabbai, A., Perpoint, B., Salque, J. R., Laurent, H., Decousus, H., and Boucheron, S.: Acro-osteolysis in leprosy. Apropos of 19 personal cases (Article in French). *Rev Rhum Mal Osteoartic 49*(2):111–119, 1982.

Radin, E. L., Parker, H. C., and Paul I. L.: Pattern of degenerative arthritis. Preferential involvement of distal finger joints. *Lancet 1*:377–379, 1971.

Raisz, L. G.: Osteoporosis. *Journal of the American Geriatrics Society 30*(2):127–137, 1982.

___: Physiology and pathophysiology of bone remodeling. *Clinical Chemistry 45*(8–pt. 2):1353–8, 1999.

Rasore-Quartino, A., Vignola, G., and Camera, G.: Hereditary enlarged parietal foramina (foramina parietalia permagna): Prenatal diagnosis, evolution and family study. *Pathologica, 77*(1050):449–455, 1985.

Rasool, M. N., and Govender, S.: The skeletal manifestations of congenital syphilis. A review of 197 cases. *Journal of Bone and Joint Surgery 71–B*(5):752–755, 1989.

Rau, R. K., and Sivasubrahmanian, D.: Supracondyloid process. *Journal of Anatomy 65*:392–394, 1931.

Ravichandran, G.: Upper lumbar spondylolysis. *International Orthopedics 5*:31–35, 1981.

Reeves, R. J., and Pederson, R: Fungous infection of bones. *Radiology 62*:55, 1954.

Reichart, P.: Facial and oral manifestations in leprosy. Anevaluation of seventy cases. *Oral Surgery Oral Medicine Oral Pathology 31*(3):385–399, 1976.

Reid, D. J., and Dean, M. C.: Brief Communication: Timing of linear enamel hypoplasia on human anterior teeth. *American Journal of Physical Anthropology 113*:135–139, 2000.

Resnick, D.: *Diagnosis of bone and joint disorders,* 4th ed. Oxford, UK: Elsevier Science, 2002.

Resnick, D., and Greenway, G.: Distal femoral cortical defects, irregularities, and excavations. *Radiology 143*(2):345–354, 1982.

Resnick, D., and Niwayama, G.: Radiographic and pathologic features of spinal involvement in diffuse idiopathic skeletal hyperostosis (DISH). *Radiology 119*(3):559–368, 1976.

Resnick, D., and Niwayama, G.: Intravertebral disk herniations: cartilaginous (Schmorl's) nodes. *Radiology 126*(1):57–65, 1978.

Resnick, D., and Niwayama, G.: *Diagnosis of bone and joint disorders.* Philadelphia: Saunders, 1981.

Resnick, D., and Niwayama, G.: Anatomy of individual joints. In Resnick, D. and Niwayama, G. (Eds.): *Diagnosis of bone and joint disorders.* Philadelphia: Saunders, 1981.

Resnick, D., and Niwayama, G.: *Diagnosis of bone and joint disorders,* 2nd ed. Philadelphia: Saunders, 1988.

Resnick, D., Niwayama, G., and Goergen, T. G.: Comparison of radiographic abnormalities of the sacroiliac joint in degenerative disease and ankylosing spondylitis. *American Journal of Roentgenology* Feb; *128*(2):189–196, 1977.

Restrepo, S., Palacios, E., and Rojas, R.: Eagle's syndrome. *Ear Nose Throat Journal 81*(10):700–701, 2002.

Rhine, S.: Non-metric Skull racing. In Gill, G. W. and Rhine, S. (Eds): *Skeletal attribution of race.* Maxwell Museum of Anthropology, Anthropological Papers No. 4:9–21, 1990.

Richards, L. C.: Temporomandibular joint morphology in two Australian aboriginal populations. *Journal Dental Research 66*:1602–1607, 1987.

___: Degenerative changes in the temporomandibular joint in two Australian aboriginal populations. *Journal Dental Research 67*:1529–1533, 1988.

Richards, L. C., and Brown, T.: Dental attrition and degenerative arthritis of the temporomandibular joint. *Journal Oral Rehabilitation 8*:293–307, 1981.

Richardson, E. G.: Tarsal Coalition. In M. S. Myerson, *Foot and ankle disorders,* Volume 1. Philadelphia: W. B Saunders Company, pp. 729–748, 2000.

Riepert, T., Drechsler, T., Urban, R., Schild, H., and Mattern, R.: The incidence, age dependence and sex distribution of the calcaneal spur. An analysis of its x-ray morphology in 1027 patients of the central European population (article in German; abstract in English). *Fortschritte auf dem Gebiete der Rontgenstrahlen und der Neuen Bildgebenden Verfahren* (Stuttgart) 162(6):502–505, 1995.

Rifkinson-Mann, S.: Cranial surgery in ancient Peru. *Neurosurgery 23*(4):411–416, 1988.

Riseborough, E, J.: Scoliosis in adults. *Current Practice in Ortopaedic Surgery 7*:36–55, 1977.

Ritschl, P., Hajek, P. C., and Pechmann, U.: Fibrous metaphyseal defects. Magnetic resonance imaging appearances. *Skeletal Radiology 18*:253–259, 1989.

Ritschl, P., Karnel, F., and Hajek, P.: Fibrous metaphyseal defects - determination of their origin and natural history using a radiomorphological study. *Skeletal Radiology 17*:8–15, 1988.

Robb, J. E.: The interpretation of skeletal muscle sites: A statistical approach. *International Journal of Osteoarchaeology 8*:363–377, 1998.

Robb, J., Bigazzi, R., Lazzarini, L., Scarsini, C., and Sonego, F.: Social "status" and biological "status": A comparison of grave goods and skeletal indicators from Pontecagnano. *American Journal of Physical Anthropology, 115*(3): 213–222, 2001.

Robbins, S. L.: *Pathology,* 3rd ed. Philadelphia: Saunders, 1968.

Robinson, H. B. G., and Miller, A.S.: *Color atlas of pathology,* 4th ed. Philadelphia: J. B. Lippincott, 1983.

Roberts, C. A.: Pers. Comm., Calvin Wells Laboratory, England, 1989.

Roberts, C. A., Boylston, A., Buckely, L., Chamberlain, A. C., and Murphy, E. M.: Rib lesions and tuberculosis: The palaeopathological evidence. *Tubercle Lung Disease 79*:55–60, 1996.

Roberts, C., Lucy, D., and Manchester, K.: Inflammatory lesions of ribs: An analysis of the Terry Collection. *American Journal of Physical Anthropology 95*:169–182, 1994.

Roche, M. B.: The pathology of neural-arch defects. *Journal of Bone and Joint Surgery, 31–A,* 529–537, 1949.

Roche, M. B., and Rowe, G. G.: The incidence of separate neural arch and coincident bone variation. *Anatomical Record, 109,* 233–252, 1953.

Rockwood, C. A.: *Fractures in children,* 4th ed. Philadelphia: Saunders, 1989.

Rockwood, C. A., and Green, D. P. (Eds.): *Fractures in adults.* Philadelphia: Lippincott, vols 1–3, 1975.

Rockwood, C. A., and Matsen, F. A. III.: *The shoulder,* 2nd ed., Vols. 1 and 2. Philadelphia: W.B. Saunders, 1998.

Rogers, J., Shepstone, L., and Dieppe, P.: Bone formers: Osteophyte and enthesophyte formation are positively associated. *Annals Rhuematic Disease 56*(2):85–90, 1997.

Rogers, L. F.: *Radiology of skeletal trauma.* New York: Churchill Livingstone, Vols 1–2, 1982.

Rogers, N. L., Flournoy, L. E., and McCormick, W. F.: The rhomboid fossa of the clavicle as a sex and age estimator. *Journal of Forensic Science 45*(1):61–67, 2000.

Rogers, J. and Waldron, T.: *A field guide to joint disease in archaeology.* New York: Wiley, 1995.

Rogers, S. and Merbs, C. F.: *Trehined skulls* (42 Slide set). San Diego Museum of Man, 1980.

Rom, W. N., Garay, S. M. and Bloom, B. R.: *Tuberculosis,* 2nd ed. Philadelphia: Lippincott Williams and Wilkins, 2004.

Rosa, Arnay-de-la M., Gonzalez-Reimers, E., Castilla-Garcia, A. and Santolaria-Fernandez, F.: Radiopaque transverse lines (Harris lines) in the prehispanic population of El Hierro (Canary Island). *Anthropologischer Anzeiger 52*(1): 53–57, 1994.

Rosenberg, N. J., Bargar, W. L., and Friedman, B.: The incidence of spondylolysis and spondylolisthesis in nonambulatory patients. *Spine 6*:35–38, 1981.

Roser, L. A., and Clawson, D. K.: Football injuries in the very young athlete. *Clinical Orthopedics and Related Research 69*:219223, 1970.

Rothschild, B.: Porotic hyperostosis as a marker of health and nutritional conditions. *American Journal of Human Biology 13*(6):709–717, 2001.

Rothchild, B., and Rothchild, C.: Treponemal disease revisited: Skeletal discriminators of yaws, bejel and venereal syphilis. *Clinical Infectious Diseases 20*(5):1402–8, 1995.

Rowe, G. G., and Roche, M. B.: The etiology of separate neural arch. *Journal of Bone and Joint Surgery 35–A:* 102–110, 1953.

Rudolph, A. H.: Syphilis. In Hoeprich, P. D. and Jordan, M. C. (Eds.): *Infectious diseases: A modern treatise of infectious processes,* 4th Ed. Philadelphia: Lippincott, pp. 666–684, 1989.

Ruge, D., and Wiltse, L. L.: *Spinal disorder: Diagnosis and treatment.* Philadelphia: Lea and Febiger, 1977.

Russell, R.G.G.: Paget's Disease. In Nordin, B.E.C (Ed.).: *Metabolic bone and stone disease.* London: Churchill Livingstone, pp. 190–233, 1984.

Ryan, D. E.: Painful temporomandibular joint. In McCarty, D. J. (Ed.): *Arthritis and allied conditions: A textbook of rheumatology,* 11th ed. Philadelphia: Lea and Febiger, 1989.

Sager, P.: *Spondylosis cervicalis: A pathological and osteoarchaeological study.* Copenhagen: Munksgaard, 1969.

Saifuddin, A., White, J., Tucker, S., and Taylor, A. B.: Orientation of lumbar pars defects: Implications for radiological detection and surgical management. *Journal of Bone and Joint Surgery 80-B*(2):208–211, 1998.

Saini, T. S., Kharat, D. U., and Mokeen, S.: Prevalence of shovel-shaped incisors in Saudi Arabian dental patients. *Oral Surgery Oral Medicine Oral Pathology 70*(4):540–544, 1990.

Salter, R. B.: *Textbook of disorders and injuries of the musculo-skeletal system.* Baltimore: Williams and Wilkins, 1970.

Saluja, P. G.: The incidence of spinal bifida occulta in a historic and a modern London population. *Journal of Anatomy 158*:91–93, 1988.

Saluja, P. G., Fitzpatrick, F., Bruce, M. and Cross, J.: Schmorl's nodes (intervertebral ruinations of intervertebral disc tissue) in two historic British populations. *Journal of Anatomy 145*:87–96, 1986.

Salvadei, L., Ricci, F., and Manzi, G.: Porotic hyerostosis as a marker of health and nutritional conditions during childhood: studies at the transition between Imperial Rome and the Early Middle Ages. *American Journal of Human Biology 13*(6): 709–717, 2001.

Sammarco, G. J. and Cooper, P. S.: *Foot and ankle manual,* 2nd ed. Philadelphia: Williams and Wilkins, 1998.

Sammarco, V.J.: Os Acromiale: Frequency, anatomy and clinical implications. *Journal of Bone and Joint Surgery 82-A*:394–400, 2000.

Sanan, A., and Haines, S. J.: Repairing holes in the head: A history of cranioplasty. *Neurosurgery 40*(3):588–603, 1997.

Santos, A. L., and Roberts, C. A.: Anatomy of a serial killer: Differential diagnosis based on rib lesions from the Coimbra Identified Skeletal Collection. *American Journal of Physical Anthropology [Supplement] 32*:130 [abstract], 2001.

Sarnat, B. G., and Schour, I.: Enamel hypoplasia (chronic enamel aplasia) in relation to systemic disease: A chronologic, morphologic and etiologic classification. *Journal of American Dental Association 28*:1989–2000, 1941.

Sarrafian, S. K. *Anatomy of the foot and ankle: Descriptive, topographic, functional* (2nd edition). Philadelphia: J. B. Lippincott, 1993.

Saunders, S. R.: *The development and distribution of discontinuous morphological variation of the human infracranial skeleton.* Ottawa: National Museum of Man Mercury Series, 1978.

Saunders, S. R., and Mayhall, J. T.: Developmental patterns of human dental morphological traits. *Archives of Oral Biology 27*(1):45–49, 1982.

Savini, R., Martucci, E., Prosperi, P., Gusella, A., and Di Silverstre, M.: Osteoid osteoma of the spine. *Journal Ortopedic Traumatologia 14*(2):233–238, 1988.

Schaeffer, J. P. (Ed.): *Morris' human anatomy: A complete systematic treatise,* 19th ed. Philadelphia: Blakiston, 1942.

Schajowicz, F.: *Tumors and tumorlike lesions of bones and joints.* New York: Springer-Verlag, 1981.

Schajowicz, F., Sainz, M. C., and Slullitel, J. A.: Juxta-articular bone cysts (intra-osseous ganglia). *Journal of Bone and Joint Surgery 61-B*(1):107–116, 1979.

Schendel, S. A., Tessier, P. and Tulasne, J.-F.: Facial Clefting disorders and craniofacial synostoses: Skeletal considerations. In Turvey, T. A., Vig, K. W. L. and Fonseca, R. J. (Eds.) *Facial clefts and craniosynostosis: Principles and management.* Philadelphia: W. B. Saunders Company, 1996.

Schlesinger, I., and Waugh, T.: Slipped capital femoral epiphysis, unsolved adolescent hip disorder. *Orthopaedic Review 16*:33–48, 1987.

Schmorl, G., and Junghanns. H.: die gesunde und dranke Wirbelsaule in Rontgenbild (article in German). *Fortschr. Rontgenstr.,* Supplement 43, Leipzig, 1932.

____: *The human spine in health and disease,* 2nd ed. (American). New York: Grune and Stratton, 1971 (a)

____: Chapter IV, Variations and malformations of the spine: and Chapter XIII, Vertebral slipping and displacement. In *The human spine in health and disease.* Second American edition.

Edited by E. F. Besemann. New York: Grune and Strattion, 1971 (b).

Schoenecker, P. L: Slipped capital femoral epiphysis. *Orthopaedic Review 14*:289, 1985.

___: Legg-Calve-Perthes disease. *Orthopaedic Review 15*(9):561–574, 1986.

Schoeninger, M. J.: *Dietary reconstruction at Chalcatzinoo, a formative period site in Morelos, Mexico.* Technical report No. 9, Museum of Anthropology. Ann Arbor, University of Michigan, 1979.

Schultz A.H.: The fontanella metopica and its remnants in the adult skull. *American Journal of Anatomy 23*:259–271, 1918.

Schultz, M.: Diseases of the ear region in early and prehistoric populations. *Journal of Human Evolution 8*(6):575–580, 1979.

Schultz, M.: Microscopic investigation of excavated skeletal remains: A contribution to pale-opathology and forensic medicine. In Haglund, W.D. and Sorg, M.H.: *Forensic taphonomy. The postmortem fate of human remains.* Boca Raton: CRC Press, pp. 201–222, 1997.

Schultz, M.: Paleohistopathology of bone: A new approach to the study of ancient diseases. *American Journal of Physical Anthropology, 33:* 106–147, 2001.

Schultz, M.: Light microscopic analysis of Skeletal Paleopathology. In Ortner, D. J.: *Identification of pathological conditions in human skeletal remains,* 2nd edition. Amsterdam: Academic Press, pp. 73–107, 2003.

Schwarz, E.: A typical disease of the upper femoral epiphysis. In Burwell, R. B. and Harrison, M. H. M. (Eds.): *Clinical Orthopaedics and Related Research 209*:5, 1986.

Scutellari, P. N., Orzincolo, C., and Castaldi, G.: Association between diffuse idiopathic skele-tal hyperostosis and multiple myeloma. *Skeletal Radiology 24*:489–492, 1995.

Seah, Y. H.: Torus palatinus and torus mandibularis: A review of the literature. *Australian Dental Journal 40*(5):318–321, 1996.

Sefcakova, A., Strouhal, E., Nemeckova, A., Thurzo, M., and Stassikova-Stukovska, D.: Case of metastatic carcinoma from end of the 8th-Early 9th Century Slovakia. *American Journal of Physical Anthropology 116*:216–229, 2001.

Seitsalo, S., Osterman, K., Poussa, M., and Laurent, L. E.: Spondylolisthesis in children under 12 years of age: Long-term results of 56 patients treated conservatively or operatively. *Journal of Pediatric Orthopaedics 8*:516–521, 1988.

Seligsohn, R., Rippon, J. W., and Lerner, S. A.: Aspergillus terreus osteomyelitis. *Archives of Internal Medicine 137:* 918–920, 1977.

Sella, E. J., and Barrette, C.: Staging of Charcot neuroarthropathy along the medial column of the foot in the diabetic patient. *The Journal of Foot & Ankle Surgery, 1,* Vol. 38: 34–40, 1999.

Semrad, W., Vanek, I., Taborsky, J., and Urban, T.: Aneurysm of the descending aorta caus-ing destruction of vertebral bodies. *Sbornik Lekarsky* (Praha) 101(3):273–279, 2000.

Shaaban, M. M.: Trephination in ancient Egypt and the report of a new case from Dakleh Oasis. *OSSA 9/11*:135, 1982–1984.

Shah, D. S., Sanghavi, S. J., Chawda, J. D., and Shah, R. M.: Prevalence of torus palatinus and torus mandibularis in 1000 patients. *Indian Journal Dental Research 3*(4):107–110, 1992.

Shands, A. R.: *Handbook of orthopedic surgery.* St. Louis: Mosby, 1951.

Sharma, P. D., and Dawkins, R. S.: Patent foramen of Huschke and spontaneous salivary fis-tula. *Journal Laryngology Otology 98*(1):83–85, 1984.

Sharma, J. C.: Dental morphology and odontometry of the Tibetan immigrants. *American Journal of Physical Anthropology 61*(4):495–505, 1983.

Sharon, R., Weinberg, H., and Husseini, N.: An unusually high incidence of homozygous MM in ankylosing spondylitis. *Journal of Bone and Joint Surgery 67-B*(1):122–123, 1985.

Shauffer, I. A., and Collins, W. V.: The deep clavicular rhomboid fossa. Clinical significance and incidence in 10,000 routine chest photoflorograms. *Journal of American Medical Association 195*:778–779, 1966.

Shields, E. D., and Mann, R. W.: Salivary glands and human selection: A hypothesis. *Journal of Craniofacial Genetics and Developmental Biology 16*(2):126–136, 1996.

Shipman, P., Walker, A., and Bichell, D.: *The human skeleton.* Cambridge: Harvard Press, 1985.

Shore, L. R.: A report of the nature of certain bony spurs arising from the dorsal arches of the thoracic vertebrae. *Journal of Anatomy 65:*379, 1931.

Sillen, A.: *Strontium and diet at Havonim Cave,* Israel, Ph.D. Dissertation. Philadelphia, University of Pennsylvania, 1981.

Silverman, F. N., and Kuhn, J . P.: *Caffey's pediatric x-ray diagnosis: An integrated imaging approach.* Mosby: St. Louis, Vols. 1 and 2, 1993.

Simmons, E. H., and Jackson, R. P.: (1979) The management of nerve root entrapment syndromes associated with the collapsing scoliosis of idiopathic lumbar and thoracolumbar curves. *Spine 4*(6):533–541.

Simpson, W. M., and McIntosh, C. A.: Actinomycosis of the vertebrae (actinomycotic Pott's disease): report of four cases. *Archives of Surgery 14:*1166, 1927.

Singer, F. R.: *Paget's disease of bone.* New York: Plenum, 1977.

Sisk, T. D.: Fractures of lower extremity. In A. H. Crenshaw (Editor), *Campbell's operative orthopaedics,* 7th edition, Volume 3. St. Louis: C. V. Mosby Company, 1987.

Sivers, J. E., and Johnson, G. K.: Diagnosis of Eagle's syndrome. *Oral Surgery Oral Medicine Oral Pathology 59:*575–577, 1985.

Sjøvold, T.: A report on the heritability of some cranial measurements and non-metric traits. In Van Vark, G.N and Howells, W.W. (Eds.) *Multivariate statistcs in physical anthropology.* Dordrecht: D. Reidel, pp. 223–246, 1984.

Smillie, I. S.: *Osteochondritis dessicans.* Edinburgh: Livingstone, 1960.

____: *Injuries of the knee joint,* 3rd ed. Baltimore: Williams and Wilkins, 1962.

Smith, G. E.: *The archaeological survey of Nubia: Report on the human remains.* Cairo: National Printing Department, 1910.

Smith, O. C., Berryman, H. E., and Lahren, C. H.: Cranial fracture patterns and estimate of direction from low velocity gunshot wounds. *Journal of Forensic Sciences 32:*1416–1421, 1987.

Smith, O. C., Berryman, H. E., Symes, S. A., Francisco, J. T., Hnilica, V.: Atypical gunshot exit defects to the cranial vault. *Journal of Forensic Sciences 38*(2):339–343, 1993.

Smith, J. T., Skinner, S. R., and Shonnard, N. H.: Persistent synchondrosis of the second cervical vertebra simulating a hangman's fracture in a child. *Journal of Bone and Joint Surgery 75*(8):1228–1230, 1993.

Snapper, I.: *Bone diseases in medical practice.* New York: Grune and Stratton, 1957.

Snodgrass, J. J., and Galloway, A.: Utility of dorsal pits and pubic tubercle height in parity assessment. *Journal of Forensic Sciences 48*(6):1226–1130, 2003.

Snow, C. E.: *Early Hawaiians: An initial study of skeletal remains from Mokapu, Oahu.* Lexington: University Press of Kentucky, 1974.

Solomon, L. B., Ruhli, F. J., Taylor, J., Ferris, L., Pope, R., and Henneberg, M.: A dissection and computer tomograph study of tarsal coalitions in 100 cadaver feet. *Journal of Orthopaedic Research 21:* 352–358, 2003.

Spitz, D. J., and Newberg, A. H.: Imaging of stress fractures in the athlete. *Radiologic Clinics of North America 40:*313–331, 2002.

Spitz, W. U.: Injury by gunfire: Gunshot wounds. In Spitz, W. U., Fisher, R. S., Editors. *Medicolegal investigation of death: Guidelines for the application of pathology to crime investigation.* Springfield: Charles C Thomas, 1980.

Spjut, H. J., Dorfman, H. D., Fechner, R. E., and Ackerman, L. V.: *Tumors of bone and cartilage.* Washington, D. C.: Armed Forces Institute of Pathology, vol. 1, 1971.

Sponseller, P. D., Bisson, L. J., Gearhart, J. R., Jeffs, R. D., Magid, D., and Fishman, E.: The anatomy of the pelvis in the exstrophy complex. *Journal of Bone and Joint Surgery 77-A*(2):177–189, 1995.

Sponseller, P. D., Jani, M. M., Jeffs, R. D., and Gearhart, J. P.: Anterior innominate osteotomy in repair of bladder exstrophy. *Journal of Bone and Joint Surgery 83-A*(2):184–193, 2001.

Spring, D. B., Lovejoy, C. O., Bender, G. N., and Duerr, M.: The radiographic preauricular groove: its non-relationship to past parity. *American Journal of Physical Anthropology 79:*247–252, 1989.

Stafne, E. C.: Bone cavities situated near the angle of the mandible. *Journal of American Dental*

*Association 29:*1969, 1942.

Stallworthy, J.A.: A case of enlarged parietal foramina associated with metopism and irregular synostosis of the coronal suture. *Journal of Anatomy 67:*168–174, 1932.

Stanitski, C. L.: Anterior Knee Pain Syndromes in the Adolescent. *Journal of Bone and Joint Surgery 75–A*(9):1407–1416, 1993.

Steele, d. G., and Bramblett, C. A.: *The anatomy and biology of the human skeleton.* College Station: Texas A&M University Press, 1988.

Steen, S. L., and Lane, R. W.: Evaluation of Habitual Activities among Two Alaskan Eskimo Populations Based on Musculoskeletal Stress Markers. *International Journal of Osteoarchaeology 8*(5):341–353, 1998.

Steinbock, T. R.: *Paleopathological diagnosis and interpretation.* Springfield: Thomas, 1976.

Stewart, T. D.: Incidence of separate neural arch in the lumbar vertebrae of Eskimos. *American Journal of Physical Anthropology 16:*51–62, 1931.

Stewart, T. D.: The circular type of cranial deformity in the United States. *American Journal of Physical Anthropology 28:*343–351, 1941.

Stewart, T. D.: The age incidence of neural-arch defects in Alaskan natives, considered from the standpoint of etiology. *Journal of Bone and Joint Surgery 35–A:*937–950, 1953.

____: Examination of the possibility that certain skeletal characters predispose to defects in the lumbar neural arches. *Clinical Orthopaedics and Related Research 8:*44–60, 1956.

____: Distortion of the pubic symphyseal surface in females and its effect on age determination. *American Journal of Physical Anthropology 15:*9–18, 1957.

____: Stone Age Surgery. Annual Report of the Board of Regents of the Smithsonian Institution. Washington, D.C.: pp. 469–491, 1958.

____: The rate of development of vertebral osteoarthritis in American Whites and its significance in skeletal age identification. *The Leech 28:*144–151, 1958.

____: Identification of the scars of parturition in the skeletal remains of females. In: Stewart T.D. (Editor). Personal identification in mass disasters. Washington, D.C.: Smithsonian Press, 1970.

____: *The people of America.* New York: Charles Scribner, 1973.

____: Cranial dysraphism mistaken for trephination. *American Journal of Physical Anthropology 42*(3):435–437, 1975.

____: Are supra-inion depressions evidence of prophylactic trephination? *Bulletin of the History of Medicine 50:*414–434, 1976.

____: *Essentials of forensic anthropology.* Springfield: Thomas, 1979.

____: Scaphocephaly in blacks: a variant form of pathologic head deformity. *Bulletin et Memoirs de la society d'Anthropologia de Paris, t. 9,* serie 13:267–279, 1982.

Stewart, T. D., and Groome, J. R.: The African custom of tooth mutilation in America. *American Journal of Physical Anthropology 28*(1):31–42, 1968.

Stewart, T. D. and Spoehr, A.: Evidence on the paleopathology of yaws. *Bulletin of the History of Medicine 26:*538–553, 1952.

Stibbe, S. P.: Anatomical note. Skull showing perforations of the parietal bone, or enlarged parietal foramina. *Journal of Anatomy 63:*277, 1929.

Stirland, A.: Pers. Comm. England, 1989.

Stirland, A. J.: A Possible Correlation between Os Acromiale and Occupation in the Burials from Mary Rose. Proceedings of the 5th European Meeting, Sienna, Paleopathology Association, pp. 327–334, 1984.

Stirland, A.: Pre-Columbian treponematosis in Mediaeval Britain. *International Journal Osteoarchaeology 1:*39–47, 1991 (a).

Stirland, A.: Diagnosis of occupationally related paleopathology: Can it be done?. In D. J. Ortner., and A. C. Aufderheide (Eds). *Human paleopathology: Current syntheses and future Options.* Washington, D. C.: Smithsonian Institution Press, pp. 40–47, 1991 (b).

Stirland, A. J.: Musculoskeletal Evidence for activity: Problems of evaluation. *International Journal of Osteoarchaeology 8:* 354–362, 1998.

Stoller, S. M., Hekmat, F., and Kleiger, B.: A comparative study of the frequency of Anterior impingment exostoses of the ankle in dancers and nondancers. *Foot and Ankle* *4*(4):201–203, 1984.

Strassberg, M.: The epidemiology of anencephalus and spinal bifida: A review. Part I: Introduction, embryology, classification and epidemiological terms. *Spina Bifida Therapy* *4*(2):53, 1982.

Struthers, J.: Supra-condyloid process in man. *The Lancet,* Feb:1, 1873.

Stuart-Macadam, P. L.: Porotic hyperostosis: Representative of childhood condition. *American Journal of Physical Anthropology* *66*:391–398, 1985.

____: A radiographic study of porotic hyperostosis. *American Journal of Physical Anthropology* *74*(4):511–520, 1987 (a).

____: Porotic hyperostosis: New evidence to support the anemia theory. *American Journal of Physical Anthropology* *74*(4):521–526, 1987 (b).

____: Porotic hyperostosis: Relationship between orbital and vault lesions. *American Journal of Physical Anthropology* *80*:187–193, 1989.

____: Porotic hyperostosis: A new perspective. *American Journal of Physical Anthropology* *87*(1):39–47, 1992.

Suchey, J. M., Wiseley, D. V., Green, R. F., and Noguchi, T. T.: Analysis of dorsal pitting in the os pubis in an extensive sample of modern American females. *American Journal of Physical Anthropology* *51*:517–539, 1979.

Suchey, J. M., Wisely, D. V., and Katz, D. Evaluation of the Todd and McKern-Stewart methods of aging the male Os Pubis. In Reichs, K. (Ed.) *Forensic osteology: Advances in the identification of human remains.* Springfield: C. C Thomas, 1986.

Suzuki, T.: Paleopathological study on osseous syphilis in skulls of the Ainu remains. *OSSA* *9/11*:153–167, 1982–1984.

____: Paleopathological study on a case of osteosarcoma. *American Journal of Physical Anthropology* *74*:309, 1987.

Swank, s. M., and Barnes, R. A.: Osteoid osteoma in a vertebral body: Case report, *Spine* *12*(6):602, 1987.

Sweet, P. A. S., Buonocore, M. G., and Buck, I. F.: Pre-hispanic Indian dentistry. *Dental Radiography and Photography* *36*:3, 1963.

Symington, J.: On separate acromion process. *Journal of Anatomy and Physiology* *34*:287, 1900.

Symmers, W. St.C.: A skull with enormous parietal foramina. *Anatomy and Physiology* *29*:329–330, 1895.

Tague, R. G.: Morphology of the pubis and preauricular area in relation to parity and age at death in Macaca mulatta. *American Journal of Physical Anthropology* *82*(4):517–525, 1990.

Takada, Y., Ishikura, R., Ando, K., Morikawa, T., and Nakao, N.: Imaging findings of elongated styloid process syndrome (Eagle's syndrome): Report of two cases (Article in Japanese). *Nippon Igaku Hoshasen Gakkai Zasshi* *63*(1):56–58, 2003.

Taylor, J. A. M. and Resnick, D.: *Skeletal imaging: Atlas of the spine and extremeties.* Philadelphia: W. B. Saunders Company, 2000.

Teele, D. W., Klein, J. O., Rosner, B. A.: Otitis media with effusion during the first three years of life and development of speech and language. *Pediatrics* *74*:282–287, 1984.

Tehranzadeh, J., Andrews, C., and Wong, E.: Lumbar spine imaging normal variants, imaging pitfalls, and artifacts. *Radiologic Clinics of North America* *38*:1207–1253, 2000.

Terry, R. J.: A study of the supracondyloid process in the living. *American Journal of Physical Anthropology* *4*:129, 1921.

____: New data on the incidence of the supracondyloid variation. *American Journal of Physical Anthropology* *9*:265–270, 1926.

____: On the racial distribution of the supracondyloid variation. *American Journal of Physical Anthropology* *14*:459–462, 1930.

____: Osteology. In Jackson, C. M. (Ed.): *Morris' human anatomy.* Philadelphia: Blakiston's, 1933.

Tessier, P.: Anatomical classification of facial, cranio-facial and laterofacial clefts. *Journal of Maxillofacial Surgery* *4*:69–92, 1976.

Testut, L.: *Traite' d' anatomie humaine I.* Poireir & Charphy, Paris, 1911.

Thawley, S. E., Panje, W. R., Batsakis, J. G., and Lindberg, R. D.: *Comprehensive management of head and neck tumors.* Philadelphia: Saunders, vol. 2, 1987.

Thieme, F. P.: *Lumbar breakdown caused by erect posture in man: With emphasis on spondylolisthesis and herniated intervertebral discs.* Anthropological Papers, Museum of Anthropology. Ann Arbor, University of Michigan Press, No. 4, 1950.

Thijn, C. J. P. and Steensma, J. Y.: *Tuberculosis of the skeleton. Focus on radiology.* New York: Springer-Verlag, 1990.

Thomas, C. L. (Ed.): *Taber's cyclopedic medical dictionary,* 15th ed. Philadelphia: Davis, 1985.

Thomas, J. L., Christensen, J. C., Kravitz, S. R., Mendicino, R. W., Schuberth, J. M., Vanore, J. V., Weil, L. S., Zlotoff, H. J., and Couture, S. D.: The diagnosis and treatment of heel pain. *Journal of Foot and Ankle Surgery, 5,* Vol. 40:329–340, 2001.

Thot, B., Revel, S., Mohandas, R., Rao, A. V., and Kumar, A.: Eagle's Syndrome. Anatomy of the styloid process. *Indian Journal Dental Research 11*(2):65–70, 2000.

Tkocz, I., and Bierring, F.: A medieval case of metastasizing carcinoma with osteosclerotic bone lesions. *American Journal of Physical Anthropology 65:*373–380, 1984.

Tod, P. A., and Yelland, J. D. N.: Craniostenosis. *Clinical Radiology 22:*472–486, 1971.

Todd, T. W.: "Cervical" rib: Factors controlling its presence and its size. Its bearing on the morphology and development of the shoulder. *Journal of Anatomy 46:*244–288, 1912.

Todd, T. W., and McCally, W. C.: Defects of the patellar border. *Annals of Surgery 74:*775–782, 1921.

Tol, J. L., Slim, E., van Soest, A. J. van Dijk, C. N.: The relationship of the kicking action in soccer and anterior ankle impingement syndrome. A biomechanical analysis. *American Journal of Sports Medicine 30*(1):45–50, 2002.

Tolman, D.E., and Stafne, E.C.: Developmental bone defects of the mandible. *Oral Surgery, Oral Medicine, Oral Pathology 24*(4):488–490, 1967.

Toohey, J. S.: Skeletal presentation of congenital syphilis: case report and review of the literature. *Journal Pediatric Orthopaedics 5*(1):104–106, 1985.

Torgersen, J.H.: A radiological study of the metopic suture. *Acta Radiologica 33:*1–11, 1950.

____: The developmental genetics and evolutionary meaning of the metopic suture. *American Journal of Physical Anthropology 9:*98–102, 193–210, 1951.

Tosi, L.: Pers. Comm., Children's Hospital National Medical Center, 1989.

Trotter, M.: Septal apertures in the humerus of American white and Negro. *American Journal of Physical Anthropology 19:*213–227, 1934.

Troup, J. D. G.: Mechanical Factors in Spondylolisthesis and Spondylolysis. *Clinical Orthopaedics and Related Research 117:*59–67, 1976.

Trueta, J.: *Studies of the development and decay of the human frame.* Philadelphia: Saunders, 1968.

Tsai, S. J. and King, N. M.: A catalogue of anomalies and traits of the permanent dentition of southern Chinese. *Journal Clinical Pediatric Dentistry 22*(3):185–194, 1998.

Tuli, S. M.: *Tuberculosis of the spine.* National Library of Medicine. Preur Printing Press, 1975.

Turk, J. L.: Syphilitic caries of the skull–the changing face of medicine. *Journal Royal Society Medicine, 88*(3): 146–148, 1995.

Turlik, M. A.: Seronegative arthritis as a cause of heel pain. *Clinics in Podiatric Medicine and Surgery 7*(2):369–375, 1990.

Turvey, T. A., Vig, K. W. L., and Fonseca, R. J.: *Facial clefts and craniosynostosis: Principles and management.* Philadelphia: W. B. Saunders Company, 1996.

Twomey, L. T. and Taylor, J. R.: Age changes in lumbar vertebrae and intervertebral discs. *Clinical Orthopaedics and Related Research 224:*97–104, 1988.

Tyson, R., and Dyer, E. S. (Eds.): *Catalogue of the Hrdlicka Paleopathology Collection.* San Diego: San Diego Museum of Man, 1980.

Ubelaker, D. H.: *Human skeletal remains: Excavation, analysis, interpretation,* 3rd ed. Washington: Taraxacum, 1999.

____: The development of American paleopathology. In Spencer, F. (Ed.): *A History of American Physical Anthropology 1930-1980.* New York: Academic, 1982.

Uemura, S., Fujishita, M., and Fuchihata, H.: Radiographic interpretation of so-called developmental defect of mandible. *Oral Surgery Oral Medicine Oral Pathology 41*(1):120–128, 1976.

Umans, J.: Ankle inpingement syndromes. *Seminar Musculoskeletal Radiology 6*(2): 133–139, 2003.

Utsinger, P. D.: Diffuse idiopathic skeletal hyperostosis (DISH, ankylosing hyperostosis). In Moskowitz, R. W., Howell, D. S., Goldberg, V. M., and Mankin, J. J. (Eds.): *Osteoarthritis: Diagnosis and management.* Philadelphia: Saunders, 1984.

___: Diffuse idiopathic skeletal hyperostosis. *Clinics in Rheumatic Diseases 11*(2):325–351, 1985.

Vailas, J. C.: Pers. Comm., George Washington University Hospital, Washington, 1989.

Van Dijk, C. N., Wessel, R. N., Tol, J. L., and Maas, M.: Oblique radiograph for the detection of bone spurs in anterior ankle impingement. *Skeletal Radiology 31*(4):214–221, 2002.

van Saase, J. L., van Romunde, L. K., Cats, A., Vandenbroucke, J. P., Valkenburg, H. A.: Epidemiology of osteoarthritis: Zoetermeer survey. Comparison of readiological osteoarthritis in a Dutch population with that in 10 other populations. *Annals of the Rheumatic Diseases 48*(4):271–280, 1989.

Verano, J.: Pers. Comm., Smithsonian Institution, Washington, 1989.

Versfeld, G. A. and Solomon, A.: A diagnostic approach to tuberculosis of bones and joints. *Journal of Bone and Joint Surgery 64-B*(4):446–447, 1982.

Velasco-Suarex, M., Bautista Martinez, J., Garcia Oliveros, R., and Weinstein, P. R.: Archaeological origins of cranial surgery: Trephination in Mexico. *Neurosurgery 31*(2):313–318, 1992.

Vidic, B.: Incidence of torus palatinus in Yugoslav skulls. *Journal of Dental Research 45:*1511–1515, 1966.

Vincelette, P., Laurin, C. A., and Levesque, H. P.: The footballer's ankle and foot. *Canadian Medical Association Journal 107*(9):872–874, 1972.

Waddington, M. M.: *Atlas of the human skull.* Vermont: Academy Books, 1981.

Waldron, T.: Unilateral spondylolysis. *International Journal Osteoarchaeology 2:*177–181, 1992.

___: A case-referent study of spondylolysis and spina bifida and transitional vertebrae in human skeletal remains. *International Journal Osteoarchaeology 3:*55–57, 1993.

Walker, P. L.: Porotic hyperostosis in a marine-dependent California Indian population. *American Journal of Physical Anthropology 69:*345–354, 1986.

___: Cranial injuries as evidence of violence in prehistoric southern California. *American Journal of Physical Anthropology 80:*313–323, 1989.

Wang, R. G., Bingham, B., Hawke, M., Kwok, P., and Li, J. R.: Persistence of the foramen of Huschke in the adult: An osteological study. *Journal of Otolaryngology 20*(4):251–253, 1991.

Wastie, M. L.: Radiological changes in serial x-rays of the foot and tarsus in leprosy. *Clinical Radiology 26*(2):285–292, 1975.

Watt, I., and Dieppe, P.: Osteoarthritis revisited. *Skeletal Radiology 19:*1–3, 1990.

Weaver, J. K.: Bipartite patella as a cause of disability in the athlete. *American Journal of Sports Medicine 5:*137–143, 1977.

Webb, S. G.: *Paleopathology of aboriginal Australians. Health and disease across a hunter-gatherer continent.* Cambridge: Cambridge University Press, 1995.

___: Two possible cases of trephination from Australia. *American Journal of Physical Anthropology 75*(4):541–548, 1988.

Weinfeld, R. M., Olson, P. N., Maki, D. D., and Griffiths, H. J.: The prevalence of diffuse idiopathic skeletal hyperostosis (DISH) in two large American Midwest metropolitan hospital populations. *Skeletal Radiology 26:*222–225, 1997.

Weinstein, P. R., Ehni G., & Wilson, C. B.: *Lumbar spondylosis: Diagnosis, management and surgical treatment.* Chicago: Year Book Medical Publishers, 1977.

Welcker, H.: Cribra orbitalia. Ein ethnologish-diagnostisches merkmal am schadel mebruer menschenrassen. *Archives Anthropologie 17:*1–18, 1888.

Wells, C.: *Bones, bodies, and disease.* London: Thames and Hudson, 1964.

___: Pseudopathology. In Brothwell, D. R., and Sandison, A. T. (Eds.): *Diseases in antiquity. A survey of the diseases injuries and surgery of early populations.* Springfield: Thomas, pp. 5–19,

1967.

____: Osteochondritis dissecans in ancient British skeletal material. *Medical History 18;*365–369, 1974.

____: Ancient lesions of the hip joint. *Medical and Biological Illustration 26:*171–177, 1976.

Wells, C., and Woodhouse, N.: Paget's disease in an Anglo-Saxon. *Medical History 19:*396–400, 1975.

West, N.F.: The aetiology of ankylosing spondylitis. *American Rheumatological Disease 8:*143–148, 1949.

Wheat, L. J.: Histoplasmosis. In Hoeprich, P. D. and Jordan, M. C. (Eds.): *Infectious diseases: A modern treatise of infectious processes,* 4th Ed. Philadelphia: Lippincott, pp. 481–488,1989.

White, S. C. and Pharoah, M. J.: *Oral radiology: Principles and interpretation,* 4th Ed. St. Louis: Mosby, 2000.

White, T. D. and Folkens, P. A.: *Human osteology.* New York: Academic Press, 1991.

Wilczak, C. A.: Consideration of sexual dimorphism, age, and asmmetry in quantitative measurements of muscle insertion sites. *International Journal of Osteoarchaeology 8*(5): 311–325, 1998.

Wilkins, K. E.: The uniqueness of the young athlete: musculoskeletal injuries. *American Journal of Sports Medicine, 8*(5):377–382, 1980.

Williams, H. U.: The origin and antiquity of syphilis: The evidence from diseased bones. *Archives of Pathology 13:*779–814, 931–983, 1932.

Williams, P. L., and Warwick, R. (Eds.): *Gray's Anatomy,* 36th ed. Edinburgh: Churchill Livingstone, 1980.

Williams, T. G.: Hangman's fracture. *Journal of Bone and Joint Surgery 57–B*(1):82–88, 1975.

Willis, T. A.: Bachache from vertebral anomaly. *Surgery, Gynecology and Obstetrics* May:658–665, 1924.

____: The separate neural arch. *Journal of Bone and Joint Surgery 13–A:*709, 1931.

Wilson, A. K.: Roentgenological findings in bilateral symmetrical thinness of the parietal bones (senile atrophy). *American Journal of Roentgenology 51:*685, 1944.

Wilson, F. C.: Fractures and dislocations of the ankle. C. A. Rockwood and D. P. Green (Editors), *Fractures,* Volume 2, 1361–1399. Philadelphia: J. B. Lippincott Company, 1975.

Wilson, G. E.: *Fractures and their complications.* Toronto: Macmillan Company of Canada Limited, 1930.

Wiltse, L. L.: The etiology of spondylolisthesis. *American Journal of Orthopedics 44A:*539–560, 1962.

Wiltse, L. L., Widell, E. H., Jr., and Jackson, D. W.: Fatigue fracture: The basic lesion in isthmic spondylolisthesis. *Journal of Bone and Joint Surgery 57–A:*17–21, 1975.

Wiltse, L. L., Newman, P. H., Macnab, I.: Classification of Spondylolysis and Spondylolisthesis. In Urist, M. R (Editor-in Chief). *Clinical Orthopaedics and Related Research* (No. 107). Philadelphia: J. B. Lippincott Company: pp. 23–30, 1976.

Witt, C. M.: The supracondyloid process of the humerus. *Journal of the Missouri Medical Association 47:*445–446, 1950.

Wong, B. J., Cervantes, W., Doyle, K. J., Karamzadeh, A. M., Boys, P., Brauel, G., and Mushtaq, E.: Prevalence of external auditory canal exostoses in surfers. *Archives Otolaryngology Head Neck Surgery 125*(9):969–972, 1999.

Wright, V.: Osteoarthritis-epidemiology. Presented at the Conference of International Symposium on Epidemiology of Osteoarthritis. Paris, June 30, 1980.

Wroble, R.R., and Weinstein, S.L.: Histocytosis X with scoliosis and osteolysis. *Journal of Pediatrics and Orthopedics 8*(2):213–218, 1988.

Wuyts, W., Cleiren, E., Homfray, T., Rasore-Quartino, A., Vanhoenacker, F., and Van Hul W.: The ALX4 homeobox gene is mutated in patients with ossification defects of the skull (foramina parietalia permagna, OMIM 168500). *Journal of Medical Genetics, 37*(12): 916–920, 2000.

Wyman, J.: *Observations on Crania.* Boston: A. A. Kingman, 1868.

Wynne-Davies, R.: Family studies and etiology of club foot. *Journal of Medical Genetics*

2:227–232, 1965.

Wynne-Davies, R., and Scott, J. H.: Inheritance and Spondylolisthesis: a Radiographic Family Survey. *Journal of Bone and Joint Surgery 61–B*(3):301–305, 1979.

Yazici, M., Kandemir, U., Atilla, B., and Eryilmaz, M.: Rotational profile of lower extremities in bladder exstrophy patients with unapproximated pelvis: A clinical and radiologic study in childern older than 7 years. *Journal of Pediatric Orthopaedics 19*(4):531–535, 1999.

Yochum, T. R., and Rowe, L. J.: *Essentials of skeletal radiology* (2nd ed). Baltimore: Williams & Wilkins, 1996.

Youmans, G. G. P., Paterson, P. Y., and Sommers, H. M.: *The biologic and clinical basis of infectious disease.* Philadelphia: Saunders, 1980.

Younes, M., Ben Ayeche, M. L., Bejia, I., Ben Hamida, R., Dahmene, J., and Moula, T.: Tubercular abscess of the psoas without associated spinal involvement. A case report (Article in French). *Revue de Medecine Interne* (Paris) *23*(6):549–553, 2002.

Yu, W., Feng, F., Dion, E., Yang, H., Jiang, M., and Genant, H. K.: Comparison of radiography, computed tomography and magnetic resonance imaging in the detection of sacroiliitis accompanying ankylosing spondylitis. *Skeletal Radiology 27*:311–320, 1998.

Zaaijer, T.: Untersuchungen uber die form des beckens javanischen Frauen, Naturrk. Verhandel. Holland Maatsch. *Wentesch Haarlem 24*:1, 1866.

Zabek, M.: Familial incidence of foramina parietalia permagna. *Neurochirurgia* (Stuttgart), *30* (1): 25–27, 1987.

Zaino, E.: *Symmetrical osteoporosis, a sign of severe anemia in the prehistoric Pueblo Indians of the Southwest.* In Wade, W. (Ed.): Miscellaneous Papers in Paleopathology. Museum of Northern Arizona Technical Series No. 7, 1967.

Zaino, D. E., and Zaino, E. C.: Cribra orbitalia in the aborigines of Hawaii and Australia. *American Journal of Physical Anthropology 42*(1):91–93, 1975.

Zeppa, M. A., Laorr, A., Greenspan, A., McGahan, J. P., and Steinbach, L. S.: Skeletal coccidioidomycosis: Imaging findings in 19 patients. *Skeletal Radiology 25*:337–343, 1996.

Zhang, Y., Jun, J., Hiroaki, I., and Katsuya, N.: Footballer's ankle: a case report. *Chinese Medical Journal* (Beijing) *115*(6):942–943, 2002.

Zimmerman, M., and Kelley, M.: *Atlas of human paleopathology.* New York: Praeger, 1982.

Zink, A., Haas, C. J., Reischl, U., Szeimies, U., and Nerlich, A. G.: Molecular analysis of skeletal tuberculosis in an ancient Egyptian population. *Journal of Medical Microbiology 50*:355–366, 2001.

INDEX

W

Wormian bones, 54i
see also Inca bone

Y

yaws (*see* treponematosis)